BIOTECHNOLOGY, AGRICULTURE AND THE FOOD INDUSTRY

Biotechnology, Agriculture and the Food Industry

G. E. Zaikov
Editor

Nova Science Publishers, Inc.
New York

For permission to use material from this book please contact us:
Telephone 631-231-7269; Fax 631-231-8175
Web Site: http://www.novapublishers.com

LIBRARY OF CONGRESS CATALOGING-IN-PUBLICATION DATA
Available upon request.

ISBN 1-60021-040-6

Published by Nova Science Publishers, Inc. ✦New York

CONTENTS

In: Biotechnology, Agriculture and the Food Industry
Editor: G. E. Zaikov, pp. 1-5
ISBN 1-60021-040-6

Chapter 1

THREE METHODS FOR DNA FINGERPRINTING OF *BRASSICA* CROPS

Yu. V. Aniskina, V. V. Martynov, A. A. Pankin, E. E. Khavkin* and I. A. Shilov

Institute of Agricultural Biotechnology, Russian Academy of Agricultural Sciences, Timiryazevskaya ul., 42, Moscow, 127550 Russia

ABSTRACT

The accessions from six *Brassica* species (*B. rapa, B. nigra, B. oleracea, B. juncea, B. napus var. oleifera,* and *B. carinata*) were fingerprinted using three independent technologies: the RFLP analysis with the hybridization probes derived from the fragments of the candidate genes that control time to flowering, the microsatellite analysis and the analysis of dispersed repetitive sequences. All three technologies discerned *Brassica* species and subspecies and produced characteristic DNA fragments specific of A, B and C genomes. The phylogenetic trees obtained three DNA technologies compatible with the botanical and cytogenetic classification of these taxa.

Key words: *Brassica - plant genotyping – introgression - RFLP - microsatellites - dispersed repeats*

The technologies of DNA genotyping that discern the species, subspecies, cultivars and clones of economic plants rapidly become an important tool for maintaining germplasm banks, breeding new cultivars, protecting breeders' rights, and ensuring the quality of seeds and other produce. In spite of the considerable progress in these areas, much is to be done to define the possibilities and limitations of DNA technologies regarding the particular crops and the specific tasks of breeding and certification.

We used three independent technologies of DNA fingerprinting to screen the accessions from six *Brassica* species that comprise the triangle of U [1]: *B. rapa, B. nigra, B. oleracea, B. juncea, B. napus var. oleifera*, and *B. carinata*. For outgroups, we used *Sinapis* and

* Fax: 7 (095) 977-0947; E-mail: emil@iab.ac.ru

Raphanus species. Seeds were obtained from the Centre for Genetic Resources, Wageningen, The Netherlands (CGN), the Vavilov Institute of Plant Industry, St. Petersburg (VIR), and the Institute of Fodder Crops, Lugovaya station, Moscow region IFC). Genomic DNA was isolated from young leaves by a modified CTAB method.

To assess the polymorphisms of the genes that control the rate of plant development, RFLP analysis was employed. The hybridization probes were derived from the clones of *Brassica* and *Camelina* homologs of the *Arabidopsis* genes *BRASSINOSTEROID INSENSITIVE1* (the Genbank accession number AF489487), *CLAVATA1* (AY130760), *CONSTANS-LIKE1* (AY379531), *FLOWERING LOCUS C3* (AY266265), *LEAFY* (AY472112), *SUPRESSOR OF CONSTANS1* (AY507669) and *SHORT VEGETATIVE PHASE1* (AY177710). The protocol for RFLP analysis was described elsewhere [2]. The primers for microsatellite analysis were chosen from the public database [3, 4]. Microsatellites were separated in 4.5% polyacrylamide gel using an ALFexpress II analyzer (Amersham BioSciences, United Kingdom) and an ALFwin Fragment Analyzer 1.00 software. The spectra of dispersed repeats were generated using the primers designed by Rogowsky *et al.* [5]. These amplicons were separated by electrophoresis in 1.5% agarose gel containing 10 mg/ml ethidium bromide in 1x TAE buffer at 22 V/cm. To build phylogenetic trees, we used the genetic distance matrix of Nei and Li [6], the UPGMA algorithm, and the software package TREECON [7].

RFLP, microsatellite and dispersed repeat technologies each produced about 120 DNA descriptors. The phylogenetic trees built upon these data (Fig. 1) are in good correspondence and completely agree with the botanical classification of these species and cytogenetic relationships between the diploid genomes AA, BB and CC and derived allopolyploid genomes AABB, AACC and BBCC [1]. It is noteworthy that, in spite of heavy selection pressure and wide diversity, the species *B. rapa*, *B. oleracea* and *B. napus* tend to form a common cluster distinct from the cluster comprising the BB species. Similar evidence was previously reported by Lowe *et al.* [3] who compared microsatellites of the AA, CC and CC genomes (*B. rapa, B. oleracea*, and *B. napus*) with those of the BB genome (*B. nigra*) and by the researchers who used several other methods of *Brassica* fingerprinting (see [8]).

The number of DNA descriptors employed in this study considerably exceeds the number of phenotypic descriptors used currently for the registration of *Brassica* cultivars and can be further expanded to discern duplicate entries in germplasm banks, essentially derived cultivars and clones.

From the practical point, a special problem is transferability of the primers used for fingerprinting six *Brassica* species. In the case of microsatellites, four primer pairs out of 25 tested produced polymorphic spectra in all species. Fourteen more were specific for particular genomes or species. All four primers used to amplify the dispersed repeats were universal for six *Brassica* species.

By characterizing DNA fragments specific of particular genomes A, B, and C, one can develop the markers for monitoring the introgression of genome-specific germplasm in segregating progenies of the crosses between *Brassica* species. Especially important are DNA markers of the B genome, a donor of such agronomic traits as earliness and disease resistances. The RFLP spectra of the flowering-time genes (Fig. 2) comprised the restriction fragments characteristic of the A, B, and C genomes, in addition, in the patterns of *FLOWERING LOCUS C* (Fig. 2b), we find DNA fragments characteristic of spring and winter forms of *B. rapa* and *B. napus* [9]. In the cladogram of *CONSTANS* restriction fragments, we could discern Brassica forms adapted to short and long photoperiods [2].

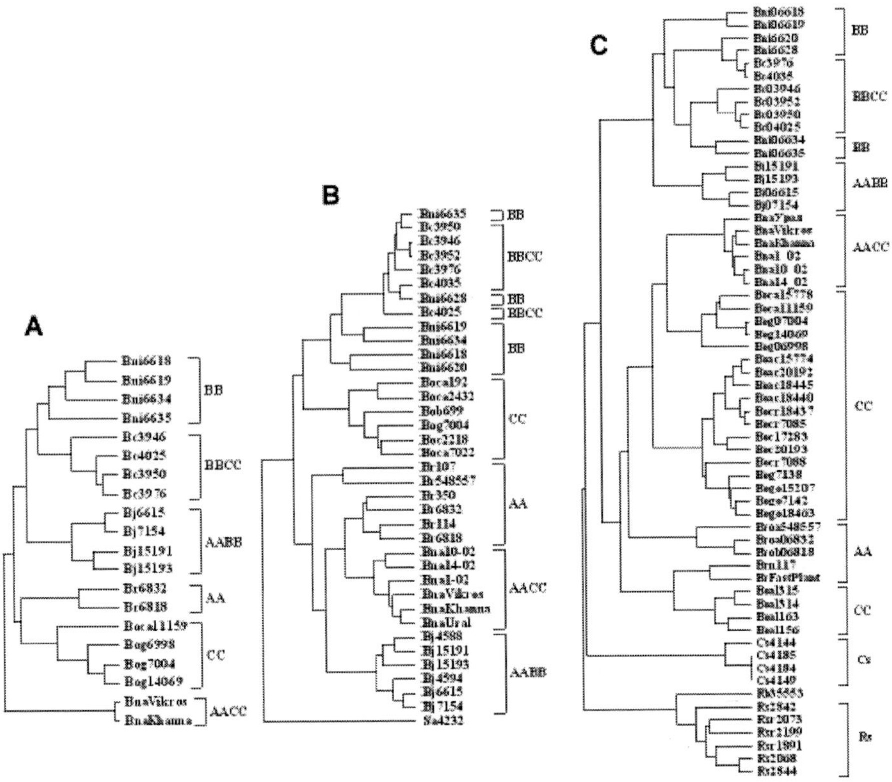

Plant accessions: genome AA: Broa - *B. rapa oleifera* annual, Brob - *B. rapa oleifera* biennial, Brn - *B. rapa* ssp. *narinosa*; genome BB: Bn - *B. nigra*; genome CC: Ba, Boac, Bob, Boca, Bocr, Bog, Bogo - *B. oleracea* var. *alboglabra*, *acephala*, *botritis*, *capitata alba*, *capitata rubrum*, *gemmiphera*, *gongyloides*, respectively; genome AABB: Bj - B. *juncea*; genome AACC: Bn - *B. napus*; genome BBCC: Bc - *B. carinata*. Outgroup species: Cs - *Camelina sativa*; Rb - *Raphanobrassica*; Rs - *Raphanus sativum*, Rsr - *Raphanus sativum* var. *radicola*; Sa - *Sinapis alba*.

Fig. 1. Cladograms of six *Brassica* species built with three sets of DNA descriptors. (A) RFLP (developmental genes as probes); (B) microsatellites; (C) dispersed repeats

By cloning and sequencing A- and B-genome specific microsatellites, we confirmed the introgression of A- and B-specific material into the AABB genome. These sequences also add new evidence to the insufficiently clear issue of polymorphism in plant microsatellites. To illustrate, when we cloned the microsatellite BRMS-042 [4] from *B. rapa*, *B. nigra* and *B. napus*, we found that A- and B-specific sequences were identical in AA and AABB genomes. Next, the genome-specific variations in the length of microsatellite sequences depended on the numbers of repeats in the major motif (AAT)n and the structure of insertions (CT)n or (CT)nTT(CT)n in the sequences that bordered the former. Finally, the allelic variations within each genome depended only on the structure of (CT)n or (CT)nTT(CT)n insertions (Fig. 3).

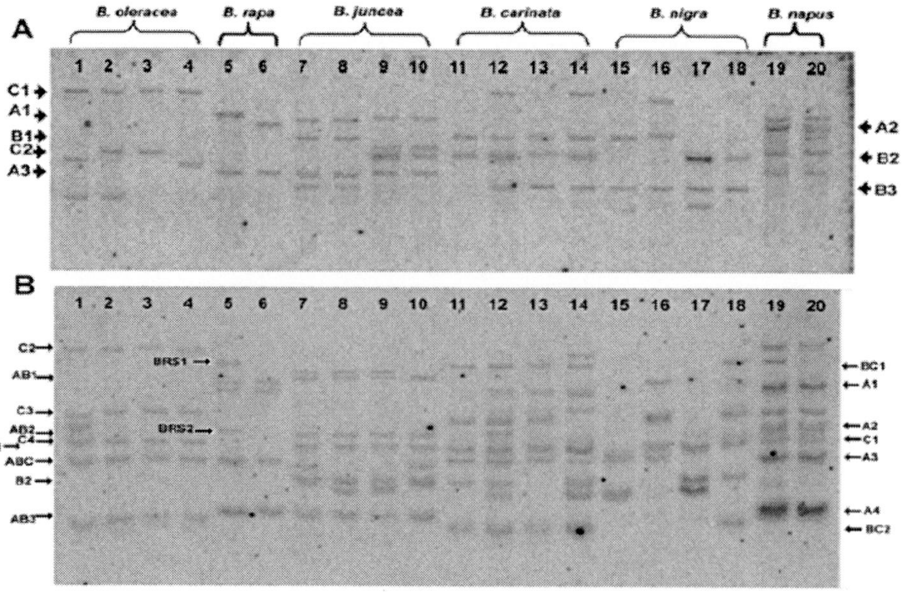

Fig. 2. RFLP spectra of two genes for time to flowering in *Brassica* species.(A) *CONSTANS*; (B) *FLOWERING LOCUS C*. Arrows indicate the restriction fragments characteristic of the genomes AA, BB, and CC. BRS - fragments characteristic of spring-annual forms of *B. rapa*.

A

AA-136
GGATCAGTTATCTGCACCACAATAATAATAATACACAAACTCTCTCTTTCTCTCTCTATCTTTCTCTCTCTATCTTTCTTATTACA
ATTCCGTTATTGGCTTTGACGTCATTATTTGAATTCTTATCCAATTCCGA
AABB-136
GGATCAGTTATCTGCACCACAATAATAATAATACACAAACTCTCTCTTTCTCTCTCTATCTTTCTCTCTCTATCTTTCTTATTACA
ATTCCGTTATTGGCTTTGACGTCATTATTTGAATTCTTATCCAATTCCGA

AA-122
GGATCAGTTATCTGCACCACAATAATAATAATACACAAACTCTCTCTTTCTCTCTCTATCTTTCTTATTACAATTCCGTTATTGGC
TTTGACGTCATTATTTGAATTCTTATCCAATTCCGA
AABB-122
GGATCAGTTATCTGCACCACAATAATAATAATACACAAACTCTCTCTTTCTCTCTCTATCTTTCTTATTACAATTCCGTTATTGGC
TTTGACGTCATTATTTGAATTCTTATCCAATTCCGA

B

BB-110 «8»
GGATCAGTTATCTGCACCACAATATAATAATACACAAACTCTCTCTCTCTCTCTTATTACAATTCGTTATTGGCTTTGACGTCATT
ATTTGAATTCTTATCCAATTCCGA
AABB-110 «8»
GGATCAGTTATCTGCACCACAATATAATAATACACAAACTCTCTCTCTCTCTCTTATTACAATTCGTTATTGGCTTTGACGTCATT
ATTTGAATTCTTATCCAATTCCGA
BB-112 «9»
GGATCAGTTATCTGCACCACAATATAATAATACACAAACTCTCTCTCTCTCTCTCTTATTACAATTCGTTATTGGCTTTGACGTCA
TTATTTGAATTCTTATCCAATTCCGA

BB-81
GGATCAGTTATCTGCACCACAATAATAATACAATTCTGTTATTGGCTTGACGTCATGATTTGAATTCTTATCCAATTCCGA
AABB-81
GGATCAGTTATCTGCACCACAATAATAATACAATTCTGTTATTGGCTTGACGTCATGATTTGAATTCTTATCCAATTCCGA

Fig. 3. Comparison of microsatellite BRMS-042 sequences cloned from *B. rapa* (AA), *B. nigra* (BB), and *B. napus* (AABB). (A) Genome A-specific, (B) genome B-specific

The dispersed repeats characteristic of the B genome have been cloned from *B. nigra, B. juncea* and *B. carinata*. In addition, we used a universal marker of the B genome based on the dispersed repeat pBNBH35 from *B. nigra* [8] and found this marker in *B. nigra, B. juncea, B. carinata*, and *B. composita*.

Another promising approach to scoring the polymorphisms of the key agronomic genes was to determine the length of the second *LEAFY* intron in *Brassica* species and subspecies; this index was shown to differ considerably within several plant families [10]. The primers for the second *LEAFY* intron were designed on the basis of the LEAFY homologs of

Brassicaceae and Solanaceae. The length of the second *LEAFY* intron in the triangle of U species varied in the range of 900 to 1000 bp, apparently due to the presence of at least three alleles.

ACKNOWLEDGEMENTS

This work was partly supported by the Russian Foundation for Basic Research, project no. 01-04-48202.

REFERENCES

[1] U, N. Genomic analysis in *Brassica* with special reference to the experimental formation of *B. napus* and peculiar mode of fertilization. *Jap. J. Bot.*, 1935, 7: 389-452.

[2] Martynov, V.V., Khavkin, E.E. Polymorphism of the *CONSTANS* gene in *Brassica* plants. Russ. *J. Plant Physiol.*, 2005, 52: 000-000.

[3] Lowe, A. J., Moule, C., Trick M., Edwards, K.J. Efficient large-scale development of microsatellites for marker and mapping applications in *Brassica* crop species. *Theor. Appl. Genet.*, 2004, 108: 1103-1112.

[4] Suwabe, K., Iketani, H. , Nunome. T. *et al.* Isolation and characterization of microsatellites in *Brassica rapa* L. *Theor. Appl. Genet.*, 2002, 104:1092–1098.

[5] Rogowsky, P.M., Shepherd, K.W., Langridge P. Polymerase chain reaction based mapping of rye involving repeated DNA sequences. *Genome*, 1992, 35: 621-626.

[6] Nei, M, Li, W.H. Mathematical model for studying genetic variation in terms of restriction endonucleases. *Proc. Nat. Acad. Sci. USA*, 1979, 76: 5269-5273.

[7] Van De Peer, Y., De Wachter, R. Treecon for Windows: A software package for the construction and drawing of evolutionary trees for the Microsoft Windows environment. *Comp. Appl. Biosci.*, 1994, 10: 569-570.

[8] Schelfhout, C. J., Snowdon, R., Cowling, W. A., Wroth, J. M. A Pcr based B-genome-specific marker in *Brassica* species. *Theor. Appl. Genet.*, 2004, 109: 917-921.

[9] Martynov, V.V., Khavkin, E.E. Rflp analysis of the *FLOWERING LOCUS C* gene in six *Brassica* species. Russ. *J. Plant Physiol.*, 2005, 52: 000-000.

[10] Frohlich, M.W., Meyerowitz, E.M. The search for homeotic gene homologs in basal angiosperms and Gnetales: a potential new source of data on the evolutionary origin of flowers. *Int. J. Plant Sci.*, 1997. 158: S131–S142.

In: Biotechnology, Agriculture and the Food Industry
Editor: G. E. Zaikov, pp. 7-16
ISBN 1-60021-040-6

Chapter 2

EXPRESSION OF MODIFIED *CRY*3A AND HYBRID *CRY*3AM-LICBM2 GENES IN TRANSGENIC POTATOES

***G. R. Salehi Jozani*[1,2,3,*], *N. Mirakhorli*[1], *I. V. Goldenkova and E. S. Piruzian*[1]**

[1]Vavilov Institute of General Genetics, Russian Academy of Sciences, 119991 Moscow, Gubkin str. 3

[2]Agricultural Biotechnology Researches Institute of Iran, 21525-1867, Karaj, Iran

[3]Biological control Research Department, Plant pests and Diseases Research Institute, Tehran 19395, Iran

ABSTRACT

The wild type *Bacillus thuringiensis cry*3a δ-endotoxin gene was cloned. On the basis of wild type *cry*3a gene sequence, modified *cry*3aM gene was designed and synthesized for effective expression in plants. The *cry*3a-*lic*BM2 and *cry*3aM-*lic*BM2 hybrid genes were constructed, in which the wild type and modified gene sequences in reading frame were fused with reporter gene, coding *Clostridium thermocellum* thermostable lichenase (*lic*BM2). The comparison expression analysis of native *cry*3a, modified *cry3a*M, hybrid *cry*3a-*lic*BM2, and *cry*3aM-*lic*BM2 genes showed that modification of the native gene sequence increased expression level of this gene in eukaryotic cells- yeasts. For Potatos transformation were constructed into plant expression vectors pC29RBCS-*leader*-*cry*3aM-*lic*BM2 and pC29RBCS-*leader*-*cry*3aM, in which sequences of *cry*3aM-*lic*BM2 and *cry*3aM were fused with a leader sequence for a pea *rbcs* gene, encoding signal peptide, which transports protein product to chloroplasts. These genes were placed under control of the light –inducible promoter for *rbcs* gene of *Arabidobsis thaliana.* Transgenic plants were received by a method of *Agrobacterium* transformation using microtubers explants. Transgenic plantlets were selected by kanamysin resistance, and PCR, lichenase activity and bioassay of neonate larvae Potato Colorodo Beetles (*Leptinotarsa decemlineata)* were conducted. Results of promoter activity assays under light induction using lichenase activity and bioassay showed high and stable expression

* Fax. 1351289, gsalehi2002@yahoo.com

of hybrid gene in transgenic plants. The presence of lichenase in the composition of hybrid proteins was shown to facilitate selection and analysis of the expression level of hybrid proteins in transgenic organisms.

Keywords: *Bacillus thuringiensis var. tenebrionis, δ-endotoxins, cry3a, Colorado potato beetle, modified gene, lichenase, licBM2.*

Introduction

Although the use of naturally occurring genes might appear to be quickest approach, many such genes will prove to be suboptimal for cloning and overexpression in heterologous systems like *E.coli*, yeasts and plants. The potential problems include high A-T contents, codon bias and complex intron/exons structures. For example, *Bt cry* genes have many splicing and polyadenylation sites, the ATTTA sequences, mRNA degradation signals and transcription termination sites as well as the codons that are rare in plants. Efforts for enhance the expression of these genes include the expression only the region of the gene encoding the insecticidal domain, modifying the UTRs, generating protein fusions, and using a variety of strong promoters. The mRNA and protein levels were eventually increased by resynthesizing the genes to be more "plant like." Increasing the G-C content of the genes may eliminate potential RNA-processing signals [2, 3, 4, 10].

We previously redesigned *cry3a* gene sequence to enhance high expression in potato cells for control of Colorado potato beetle (*Leptinotarsa decemlineata*). Then the native and modified genes were fused with the reporter gene *lic*BM2 which codes for a thermostable lichenase [5, 6, 8] and cloned into bacterial and yeast expression vectors. Using the quantitative and qualitative assays of lichenase in the hybrid proteins was assessed expression quantity of the native and modified genes in *E.coli* and yeasts cells. We reported that this modification of the coding sequence dramatically increases the expression (up to 10-fold compared to the native gene) in yeast (*Saccharomyces cerevisiae*) cells [7]. We report here that this modified form of the *cry*3aM gene is expressed at the high levels (up to 100-fold compared to the native gene) in transgenic potato plants, and that this expression confers a high level of protection from damage by Colorado potato beetles (CPB).

It was shown that the thermostability of lichenase allows one easily to test for its expression against the background of thermolabile plant enzymes. In the experiments on the inducible expression of the *cry*3aM-*lic*BM2 gene in transgenic plants, the enzymatic activity of lichenase changed under the action of light-inducible promoter small subunit ribulose-1.5-bisphosphate carboxylase gene of *Arabidobsis thaliana*. It was demonstrated that the availability of lichenase consisting of the hybrid proteins facilitates the selection and analysis of the hybrid proteins in transgenic organisms. This is very important to carry out the exploring researches.

MATERIAL AND METHODS

Strains and plasmids: Standard procedures were used for DNA manipulation [9]. The *Bt var. tenebrionis*, the *E. coli* strain XL1-Blue ("Stratagene) and BL21(DE3) ("Novagene", USA), the *Agrobacterium thumefaciens* strain ALGO were cultured and transformed as described [4] and plasmids pTZ57 R/T, pUC18, pET32b(+), pGEM-easy, pC29 were used as the cloning vehicles. *E.coli* strains used, cloning vectors and growth conditions for them were the same as previously described [9].

Gene synthesis: Synthetic gene was designed and synthesized in four fragments with length about 500 bp. Besides, in the beginning and end of each fragment were added restriction sites, which exist inside the sequence of native *cry*3a gene (*NdeI, Bsu36I, XbaI*). Oligonucleotides were prepared by the solid phase phosphoramidite triester couplings approach, using ASM-700 and ASM-800 (OOO "«Bioset", Novosibirsk, Russia) DNA synthesizer (fig. 1).

Plant transformation: For plant transformation were constructed the plant expression vectors pC29RBCS-*leader-cry*3aM-*lic*BM2 and pC29RBCS-*leader-cry*3aM (fig. 2). These vectors express foreign genes in plants under control of the promoter of a light-inducible ribulose-1.5-bisphosphate carboxylase small subunit gene of *Arabidopsis thaliana*. These constructions were cloned into the T-DNA segment between the right- and left- border region and introduced into potato tuber disks via *Agrobacterium thumefaciens* mediated transfer. The *A. thumefaciens* ALGO strain used to deliver these plasmids. The transgenic plants were rooted on kanamycin-containing media in two stages, first at 100 mg/l kanamycin and then at 200 mg/l kanamycin. Transformed plants were selected for kanamycin resistance, lichenase activity, PCR, and analyzed for resistance to damage by CPB.

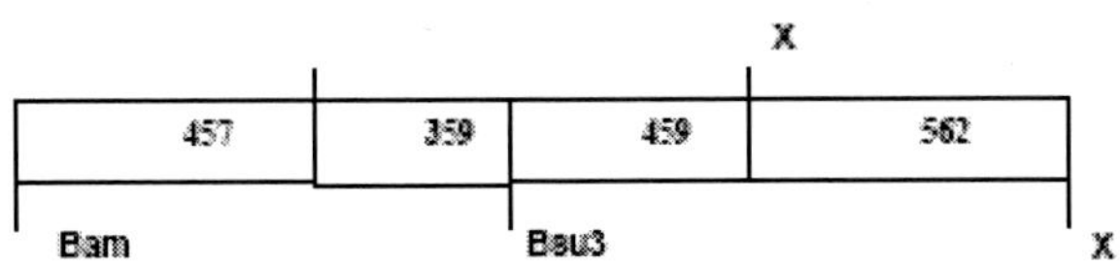

Figure 1. Scheme of modified gene (*cry*3aM) sequence, combined from four synthetic fragments

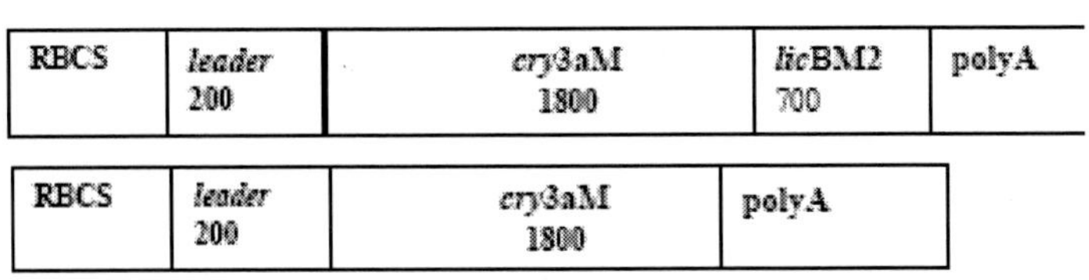

Figure 2. Vector constructs bearing the leader sequence, *cry*3aM, and *C. thermocellum lic*BM2 gene for expression in potato cells. RBCS: Light inducible promoter of small subunit of rbcs gene of *Arabidobsis thaliana*, Leader: leader sequence for pea *rbcs* gene, PolyA- Site polyadenylation

Preparation of plant protein extracts: Leaves of 1-month-old transgenic plants were ground in liquid nitrogen with a mortar and pestle. The resulting powder was extracted with 100 mM Tris-HCl pH 8.0 (1:1 w/v). The extract was centrifuged in an Eppendorf microcentrifuge for 5 min, and the supernatant obtained was incubated at 70°C for 1 h, and then centrifuged once more for 3 min. The clear supernatant obtained was used for protein determination and enzyme assay.

Protein determination and enzyme assay: Total plant protein was measured according to Bradford [1], using the BioRad dye reagent and bovine serum albumin as standard. Lichenase activity in the lysates was determined as described previously at 65°C using lichenan as substrate [5, 6]. Reducing sugars released from the substrate were determined with the 3,5 dinitrosalicylic acid (DNS) reagent according to Wood and Bhat [12]. The reaction mixture contained 200 μl of 0.5% lichenan and 100 μl of the protein sample. It was incubated for 30 min, then 1.2 ml of the dinitrosalicyl reagent was added, and the mixture was heated at 100°C for 15 min. The concentration of colored product was determined with an Ultraspec II spectrometer (LKB).

Activity staining for the enzyme in SDS-polyacrylamide gels containing lichenan was done according to Schwarz et al. (1987) with some modifications. SDS-PAGE was done either in 10% or 8-16% gradient polyacrylamide slab gels. Lichenan was added to the separating gel up to a concentration of 0.1%, prior to polymerization. Protein samples (1-10 μg) were applied to the gel after heating in 2x protein loading buffer at 100°C for 5 min. Upon completion of electrophoresis, the gels were washed twice at room temperature for 15 min in 25 mM Tris-HCl pH 8.0. The gels were next incubated at 65°C for 2-3 h in 25 mM Tris-HCl pH 8.0. Lichenase was visualized after staining the gels with Congo red (0.5 mg/ml, Sigma). The bands of active enzyme appear as clear spots since the dye binds only to unhydrolyzed lichenan.

Insect bioassay: The CPB larvae were from colonies collected as eggs from the potato fields. First instar larvae (5 per replicate) were placed in screen mesh bags within one day of eclosion (hatch), as soon as they were ready to feed and were offered no other food. Each R0 regenerant was tested with at least three replicates of first instar larvae. At least three plants were tested for each clone. A single leaf from kanamycin and lichenase activity selected plants was incubated for 4 days with 5 newly hatched neonate CPB larvae. Mean temperature for rearing and bioassays was 22°C.

Results and Discussion

Design of the cry3a coding sequence: The purpose of redesigning the cry3a gene was to create a synthetic gene (cry3aM) that would be highly expressed in plant cells and thus confer resistance to CPB larvae in transgenic potatoes. The cry3a gene has an open reading frame equivalent to a 73 kDa protein [10]. To make the cry3a gene more convenient for plant, we referred to the codon usage table of many dicot and monocot genes. In the modified gene, the cry3a sequence was modified to (1) contain codons preferred by plant genes, (2) avoid CG and TA dinucleotides at codon positions 2 and 3, (3) attain an A-T nucleotide base composition similar to that found in plants, (4) eliminate sequences that might cause mRNA destabilization and splicing, (5) minimize secondary structures hairpins, and (6) optimize the ATG flanking nucleotides for protein translation initiation. The modified gene (cry3aM) differs from the native gene in 21% of its nucleotides (table 1).

Table 1. Comparison of the native and modified *cry*3a genes

	Wild type gene (*cry*3a)	Synthetic gene (*cry*3aM)
Bases different from wild type	-	350/1812(20%)
Codons different from wild type	-	320/597 (54%)
G+C content (%)	37	49
Potential polyadenylation sites	24	0
ATTTA sequences	12	2
A+T rich regions (> 6 consecutive A and/or T)	37	0

Synthesis of modified cry3a (cry3aM) gene sequence: A new strategy was used for the chemical synthesis of the modified gene. First the gene sequence was divided into four fragments about 500 bp long (fig. 1). Then, the coding (first) strand of each of the fragments was subdivided into 50-55 bp fragments, and the corresponding oligonucleotides were synthesized by means of phosphoramidite chemistry. Oligonucleotides were prepared by the solid phase phosphoramidite trimester couplings approach, using a synthesizer. This was followed by synthesizing a set of supporting complementary oligonucleotides 30-35 bases long which were complementary to the junctions between first chain oligonucleotides, so that 15-17 5'-end bases of the supporting oligonucleotides were complementary to the corresponding 3'-end of one oligonucleotide of the first strand and the remaining 15-17 3'-end bases of the same supporting oligonucleotide were complementary to the corresponding bases of the next first strand oligonucleotide. The mixture of first strand oligonucleotides (50-55 bases long) was phosphorylated. After inactivation of the enzyme, a mixture of the supporting oligonucleotides (30-35 bases long) at the same concentration was added. The combined mixture was exposed to DNA polymerase I and the dNTPs to complete synthesis of the second strand, and then T4 DNA ligase was added to join the adjacent oligonucleotides. The assembled fragments were then subjected to PCR, the product was gel purified and cloned, either directly or by digesting with the appropriate restriction enzymes, to check the nucleotide sequence.

Cloning of the assembled fragments for complete gene construction: The complete gene sequence was achieved by cloning of all the four fragments in the correct reading frame to a pGEM vector. For obtaining plasmid pGEM-f1-f2-f3-f4cry3a (cry3aM) conducted some intermediate cloning steps. pGEM-f1-f2 was obtained by cloning of the NdeI-fragment of pGEM-f2, (containing the second fragment of the cry3aM gene) to pGEM-f1, hydrolyzed on site NdeI. For obtaining pTZ-f1-f2-f3, XbaI-BamHI fragment (f3) from pQE-81-f3 cloned to pTZ-licBM2, hydrolyzed by XbaI and BamHI endonucleases, with removing the licB sequence. After that, Bsu36I-ApaI fragment (f1-f2), was cloned to plasmid pTZ-f3, hydrolyzed by endonucleases Bsu36I and ApaI, and obtained plasmid pTZ-f1-f2-f3. Plasmid pGEM-f1-f2-f3-f4cry3a (cry3aM) was obtained by cloning of XbaI-PstI fragment (f1-f2-f3) to pGEM-f4, hydrolyzed by XbaI and PstI endonucleases (fig. 1).Such a strategy ensured the correct positioning of all oligonucleotides of the first strand, and hence the correct assembly of the target fragment and the whole gene.

Use of a new reporter system for prokaryotic and eukaryotic cells based on the thermostable lichenase from Clostridium thermocellum: Note that gene cry3a codes for a protein lacking enzyme activity, which would allow to quantitatively estimate its expression level and molecular mass of the protein product. We suggested that, to analysis of the expression efficiency of the modified cry3a gene in plant cells, thermostable lichenase can be used as a translational reporter. For this, we constructed hybrid gene cry3aM-licBM2. Lichenase is a thermostable enzyme that specifically hydrolyzes β-1,4- linkages adjacent to β-1,3- linkages in mixed-linkage β-glucans, but does not attack pure 1,3- or 1,4- linkages. Lichenase activity is not present in most bacteria, yeast, mammals, or plants, which make the enzyme a good reporter for these heterologous systems. It was shown that the thermostability of lichenase allows one easily to test for its expression against the background of thermolabile bacterial, yeast and plant enzymes. Zymogram analysis of the expression of the lichenase gene reveals the enzyme activity not only at a size corresponding to the predicted molecular mass of the enzyme, but also in a band of lower molecular mass. The high activity and thermostability of the C. thermocellum lichenase made it a good candidate for use as a reporter protein in bacterial, yeast, mammalian, and plant cells. The high specific activity of the enzyme (2000 U) makes possible qualitative and quantitative measurements even at low levels of expression. The zymogram method can be used instead of Western blotting for estimation of the expression level and size of fusion proteins.

Expression of hybrid cry3a-licBM2 gene in transgenic potatoes. For potato transformation were constructed plant expression vectors pC29RBCS-leader-cry3aM-licBM2 and pC29RBCS-leader-cry3aM, in which sequences of cry3aM-licBM2 and cry3aM were fused with leader sequence for pea rbcs gene, encoding signal peptide, which transports protein product to chloroplasts. These genes were placed under control of the light– inducible promoter for rbcs gene (small subunit gene for rib-1,5-bisphospate carboxylase) of Arabidobsis thaliana. This promoter provides expression of hybrid genes only in green tissues of plant. The combination of rbcs promoter and leader sequence is capable of expressing foreign genes at a high level in transgenic plants.

DNA isolated from leaves of all the kanamycin-resistant plants was individually analyzed by PCR for the presence of the cry3aM, licBM2 genes and leader sequence. The PCR-positive plants were further tested for the presence of the lichenase activity and were either discarded or grown to maturity inside the growth house in peat pots. Table 2 summarizes the results of our transformation experiments. Based on lichenase activity, PCR and zymogram analysis (Fig. 3, 4), fourteen primary independent transformants expressing high levels of the Bt toxin were identified.

Table 2. Summary of transformation experiments

Genotype	s-l*	s-l	s*	s
Variety	UB*	DS	UB	DS
Explant quantity	332	207	302	81
Received regenerants	150	23	59	18
Kanamycin- selected regenerants	11	6	4	5
PCR positive regenerants	8	6	-	-
Lichenase activity-selected regenerants	8	6	-	-

* UB and DS: varieties Ubiley Zhukova and Desnitsa. S-L: (*cry*3aM-*lic*BM2), S: *cry*3aM

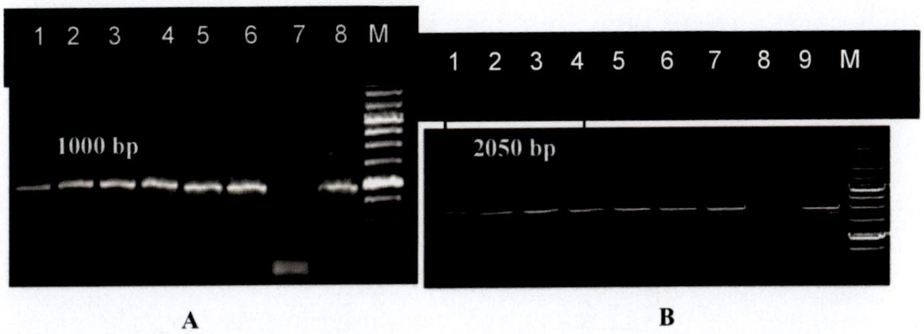

Figure 3. PCR analysis of transgenic plants. A. Using primers of 5'-end sequence of leader and 3'- end sequence of the second fragment of *cry*3aM (lanes 1-6: primary transgenic lines Leader-Cry3aM-LicBM2, lane 7: control plant (negative control), and lane 8: positive control (plasmid)). B. Using primers of 5'-end sequence of the second fragment of *cry*3aM and 3'-end sequence of *lic*BM2 (lanes 1-7: primary transgenic lines Leader-Cry3aM-LicBM2, lane 8: control plant (negative control), lane 9: positive control, M: Molecular weight marker)

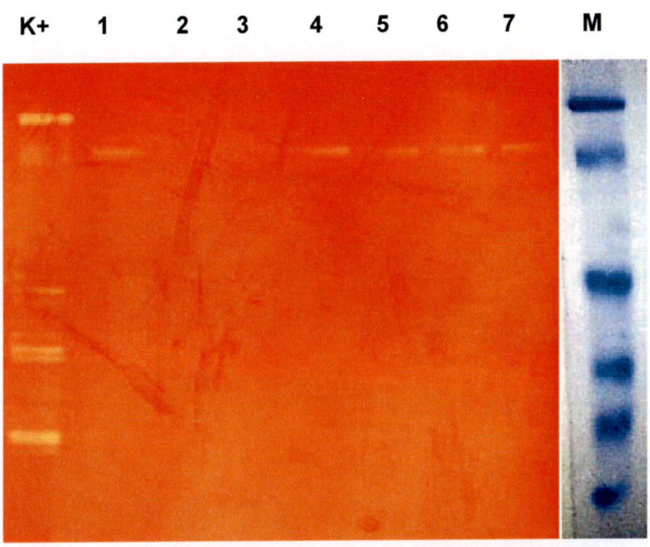

Figure 4. Zymogram of the lichenase activity in plant protein extracts detected in the presence of 0.1% lichenan as substrate. Lanes 1, 3-6: extracts of primary transgenic lines Cry3aM-LicBM2. Lane 2. non transgenic plant extract, K^+ - positive control (bacterial extract containing Cry3aM-LicBM2), 7- Marker

Before we demonstrated that method zymogram shows all of Western blot analysis characteristics, and in addition, it is very simple and inexpensive. Five independent transformants were used for bioassay on neonate larvae CPB. Using quantitative measurement of lichenase activity in structure of hybrid protein were estimated the levels of the cry3aM-licBM2 hybrid protein present in leaves of primary transgenic plants. We estimated that the hybrid protein constituted 0.05–0.1% of the total soluble protein of the plant tissue (approximately 50-100 times more than the wild type gene). Results of this experiment were summarized in table 3. Zymogram analysis indicated the presence of single band of 95 kDa, which correspond to the theoretically computed molecular masses of the hybrid protein. Using the zymograms method, we showed that lichenase contained in the hybrid proteins preserves its properties, that is, thermostability and activity, and the molecular masses of the hybrid protein could be thus inferred from the lichenase activity (fig. 4).

Kinetic analysis of light effect (inductor) on expression of *cry*3aM-*lic*BM2 hybrid gene in transgenic plants. The following experiments were conducted to determine stability of hybrid protein in primary transgenic plants. The primary transgenic plants (lines Leader-Cry3aM-LicBM2) in the age of 1 month were maintained in darkness within 7 days. Further plants were transferred to light conditions (4000-6000 luxs). Enzyme activity of lichenase, which is included in structure of hybrid protein, was measured in leave extracts of primary transgenic plants at the end of light and dark periods, and also through 6, 12 and 24 and 48 hours after transferring of plants to light conditions (fig. 5). The obtained results testified that hybrid protein (Cry3aM-LicBM2) is characterized by a high level of stability in transgenic plants. This conclusion is based on that the accumulation level of Cry3aM-LicBM2 in conditions of light absence (as inductor of rbcs promoter) within 7 days is made about 40 % of level of this protein in standard conditions of plant cultivation (16 hours - light, 8 hours - darkness). The level of enzyme activity for plants transferred to light was augmented and after 24 h light induction was reached initial level.

Table 3. Different characteristics of transgenic plants carrying *cry*3aM-*lic*BM2 hybrid gene

Independent transgenic lines	Lichenase activity (unit/mg total soluble protein)	Level of accumulation recombinant hybrid protein (%of the total soluble protein)	Mortality (%)
UB-S-L(1)	0,44 ± 0,05	0,03	37
UB-S-L(2)	1,16 ± 0,20	0,08	-
UB-S-L(3)	0,90± 0,19	0,06	55
UB-S-L(4)	0,87 ± 0,10	0,06	61
UB-S-L(5)	0,44 ± 0,05	0,03	-*
DS-S-L(1)	0,6 ± 0,10	0,04	38
UB-S-L(6)	1,10 ± 0,14	0,07	-*
UB-S-L(7)	1,20 ± 0,20	0,09	-*
DS-S-L(2)	0,34 ± 0,07	0,02	-*
DS-S-L(3)	1,25 ± 0,25	0,1	81
DS-S-L(4)	0,5 ± 0,07	0,03	-*
DS-S-L(5)	1,08 ± 0,18	0,08	-*
DS-S-L(6)	0,60 ± 0,10	0,04	-*
UB-S-L(8)	0,73 ± 0,10	0,05	-*

*. these lines were not assayed.

UB and DS: varieties Ubiley Zhukova and Desnitsa. S-L: (*cry*3aM-*lic*BM2)

Evaluation of biological activity of the hybrid protein in transgenic potatoes. Entomocidal activity of the hybrid protein in the leaves of five expressive independent transgenic plants was assayed after 5 days (table 3) through a feeding assay of neonates of CPB. Mortality ranged between 37% and 81%, and surviving insects suffered reduction in body weight ranging from 50- to 70% and failed to pupate. Results of bioassay on CPB were summarized in table 3.

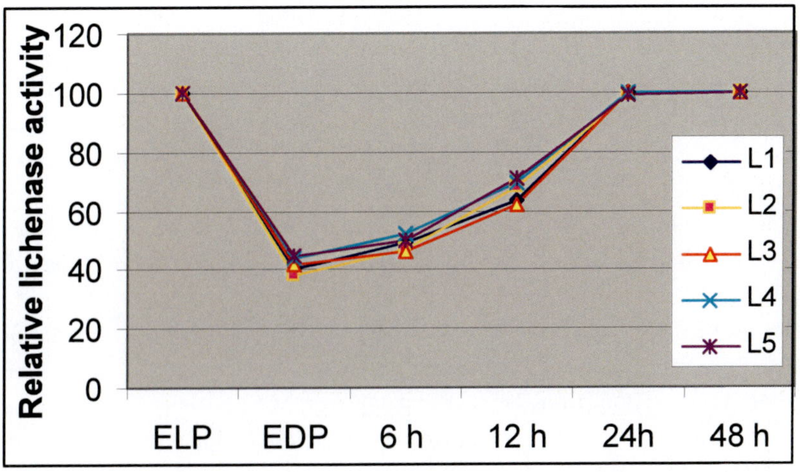

Figure 5. Kinetic analysis of light effect (inductor) on expression of *cry*3aM-*lic*BM2 hybrid gene in transgenic plants. Lichenase activity was measured at the end of light period (ELP); at the end of dark period (EDP); after 6, 12, 24, and 48 hours light induction. Ln: independent transgenic lines

Thus, the results of our experiments suggest the following conclusions. Our modification of the *cry*3a gene sequence increased the expression of this gene in eukaryotic (plant) cells. The presence of lichenase in the hybrid protein composition was shown to facilitate selection and analysis of the recombinant protein expression in transgenic organisms, which is of importance for fundamental and applied studies. Furthermore, based on the propertied of lichenase as a reporter protein (thermostability and high relative activity), this system seem promising for monitoring transgenes in agrocenoses, since it is simple, precise, inexpensive, and not time-consuming.

REFERENCES

[1] M. M. Bradford, A rapid and sensitive method for quantification of microgram quantities of protein utilizing the principle of protein-dye binding. *Anual Biochem* 72:248-254(1976).

[2] W.S. Kuo And K.F. Cha,1996, Identification of novel cry type genes fro *Bt* strains on the basis of RFLP of the PCR amplified DNA. *Appl and Envir Microbiology*, 62: 1369-1377.

[3] N. Crickmore, D.R. Zeigler, J. Feitelson, E. Schnepf, And J. Van Riel, Revision of the Nomenclature for the *Bt* pesticidal crystal proteins, *Micro and Mol Biol Reviews,* 807-813 (1998).

[4] F. J. Perlak And B. S. Terry, Genetically improved potatoes: Protection from damage by Colorado potato beetles, *Plant Molecular Biology*, 22: 313-321(1993).

[5] E.S. Piruzian, I.V. Goldenkova, K.A. Musiychuk, N.S. Kobets, I. P. Arman, I.V. Bobrysheva, I.A. Chekhuta, and D. Glazkova, A reporter system for prokaryotic and eukaryotic cells based on thermostable *Clostridium thermosellum* lichenase. *Mol.Genet. Genomics*: 266: 778-786 (2002).

[6] E.S. Piruzian, B. Monzavi-Karbassi, N.S. Darbinian, I.V. Goldenkova, N.S. Kobets, and A.V. Mochulsky, The use of a thermoactive- glucanase gene from *Clostridium thermocellum* as a reporter gene in plants. *Mol Gen Genet* 257:561-567(1998).

[7] G. R Salehi Jozani, R. A. Komakhin, and E. S. Piruzian, Comparative study of the expression of the native, modified, and hybrid *cry*3a genes of *bacillus thuringiensis* in prokaryotic and eukaryotic cells, *Genetika* , 41(2): 171-177 (2005).

[8] R. A. Komakhin, I. A. Abdeeva, G. R. Salehi Jozani, I. V. Goldenkova, and A. A. Zhuchenko, Thermostable lichenase as a translational reporter, *Genetika*, 41(1): 40-47 (2005).

[9] J. Sambrook, E.F. Fritsch, and T. Maniatis, *Molecular cloning: a laboratory manual* (2nd). Cold Spring Harbor Laboratory Press, Cold Spring Harbor, N.Y (1989).

[10] V. Secar, D. V. Thompson, M. J. Maroney and M. J. Adang, Molecular cloning and characterization of the ICP gene of *Bt var tenebrionis*, *Proc. Natl. Acad. Sci.* 84: 7036-79040 (1987).

[11] R. W. Teather, and P. J. Wood, Use of Congo Red-polysaccharide interactions in enumeration and characterisation of cellulolytic bacteria from the bovine rumen. *Appl Envi Microbiol* 43:777-780 (1982).

[12] T. M. Wood, and K. M. Bhat, Methods for measuring cellulase activities. *Methods Enzym.* 160:87-112 (1988).

In: Biotechnology, Agriculture and the Food Industry
Editor: G. E. Zaikov, pp. 17-20 ISBN 1-60021-040-6

Chapter 3

CURRENT STATUS ON GENETIC VARIATION OF POLISH HUCUL HORSES IN THE 12 MICROSATELLITE LOCI

Tomasz Ząbek, Tadeusz Rychlik and Ewa Słota
National Research Institute of Animal Production,
Krakowska 1, 32-083 Balice, Poland

ABSTRACT

Evaluation of genetic variation of Hucul primitive horses was conducted with the use of 12 microsatellite typing data obtained during routine parentage testing. Hucul horses belong to the group of endangered primitive horses and extensive efforts were made to enlarge the breeding stock of them in the past. Obtained marker data characterises current genetic status of Hucul horses on a group of 58 most unrelated individuals. Fis values revealled no loss of genetic variation in the 12 microsatellite loci. Studied population was in the state of H-W. The levels of presented variation of investigated population were comparable to other common Polish and foreign horse breeds, as was indicated by the values of heterozygosity and effective allele number. This data are in agreement with the substantial increase in population size of these primitive horses. However high values of heterozygosity may be the consequence of the use of crossbreeding matings in the Hucul population to reduce inbreeding in the past. Thus, the comparable level of their genetic variation to other horse breeds would be the result of the influence of other breeds on Hucul gene pool.

INTRODUCTION

Hucul horses originate from east Carpathians and were developed under influence of primitive groups of horses from Asia and also Tharpans and Przewalski horses in noric and oriental types. At the beginning Hucul horses were used as pack and saddle horses, and later also as carriage horses. Like other groups of primitive breeds they are characterised with great hardiness and good feed conversion. Poland has 60% of world Hucul population, comprising approximate 1500 horses (nearly 450 brood mares and 70 stallions).

In the XX century after II world war Hucul population was greatly reduced and endangered with extinction. A conservation program developed to rescue Polish Hucul population assumed pure breeding with no substantial increase of matings of close relatives. Program related on the use of 7 stallions derived from different male lines and breeding stock from Rumanian, Slovak and Czech studs (Jackowski 2002). Despite the broad efforts to maintain pure breeding stock of Hucul horses, stallions and mares of other breeds are also used in matings.

In order to evaluate the genetic status of current Hucul population their genetic variation were investigated in loci of 12 short tandem repeat (STR) systems for horse parentage testing.

MATERIALS AND METHODS

Material consisted of 58 Hucul horses being progeny of 41 fathers and 28 mothers. DNA was prepared from hair follicles with proteinase K digestion followed by ethanol purification (Kawasaki 1990). DNA was the template for two multiplex PCR reactions (8 and 4 plex). Amplified microsatellites are listed in table 1. PCR conditions were described in manufactures protocol for commercially available kit for horse parentage genotyping (Applied Biosystems, Foster City). Amplification products were separated in 4% acrylamide gel during electrophorese on the 377 laser sequencer (Applied Biosystems). Automated DNA sizing was performed with the use of GeneScan 2.1 software and microsatellite genotypes were established in Genotyper 2.0 software. Biometric calculations comprised estimation of allelic frequency and evaluation of Hardy-Weinberg proportions (the exact test) implemented in TFPGA program (Miller 1997), and estimation of Fixation coefficient - Fis (Weir and Cockerham 1984), expected heterozygosity - $\overline{H}$ (Nei 1978), and effective allele number - ne in MSA program (Dieringer and Schlötterer 2002).

RESULTS AND DISCUSSION

Exact test for calculation of Hardy-Weinberg proportions showed no deviation from the genetic equilibrium state in all investigated loci except marker HTG6. This was confirmed by obtained mean Fis values close to 0 indicating no loss of genetic variation in all loci with exception to locus HTG6. In this locus significant negative Fis value was observed indicating increase of homozygosity, due to uneven distribution of alleles in pool of HTG6 variants with domination of 100 bp allele (data not shown).

In Hucul population values of heterozygosity and effective allele number for all microsatellites without HTG6 (tab.1) are typical in different horse populations confirming their great potential for horse parentage control and population studies (Bowling et al. 1997, Hamanova et al. 2001, Ząbek et al. 2003). Mean number of microsatellite variants, mean heterozygosity values together with averaged effective number of alleles in Hucul horses (tab.1) express comparable amount of variation of Hucul gene pool to the variation of big populations of common horse breeds and utility types of horses. Higher heterozygosity values

($\overline{H}$ > 0,7) for Hucul horses were also typical for a warm blood horses widely crossed with Thoroughbreds, like Oldenburg horses, Polish Silesian horses (Ząbek et al. 2003) and Czech warm blood horses (Hamanova et al. 2001). In contrast lower $\overline{H}$ values are typical for breeds with closed studbook like Thoroughbreds ($\overline{H}$ < 0,7) (Cannon et al. 2000, Ząbek et al. 2003) expressing their genetic consolidation as a result of directed selection on racing ability (Bailey 1998). Lower heterozygosity values are also characteristic for small populations of endangered primitive horses like le cheval Canadien (Behara et al. 1998).

Table 1. Fis coefficient, allele number (na), effective allele number (ne), heterozygosity ($\overline{H}$)

Locus	Fis	na	ne	$\overline{H}$
AHT4	-0,0502	8	4,09	0,762
AHT5	0,0085	7	3,29	0,702
ASB2	-0,0061	9	5,05	0,809
HMS2	-0,0772	7	4,26	0,772
HMS3	0,0319	6	5,04	0,808
HMS6	-0,0597	7	3,73	0,739
HMS7	-0,105	6	3,19	0,693
HTG10	0,0492	8	3,74	0,739
HTG4	0,0882	6	3,13	0,687
HTG6	-0,1499	7	1,60	0,378
HTG7	0,0435	4	2,36	0,582
VHL20	0,0274	10	7,61	0,876
Mean	-0,0166	7,08	3,92	0,712

Polish population of Hucul horses is still growing under assumption of purebred matings. However a number of unrecorded matings and parentage failures were noted and efforts are in progress to correct failures in studbook data and to establish breed status for a number of individuals. So such a great variation observed in the pool of microsatellite variants in studied Hucul population may be a consequence of participating of other horse breeds in matings with Hucul horses which sometimes were needful to avoid matings of close relatives. The interbreed relationships between Hucul horses and other common and primitive horse breeds are the subject of the ongoing study and will further elucidate the degree of genetic distinction of current Hucul gene pool.

REFERENCES

[1] Bailey E. (1998). Odds on the FAST gene. *Genome research*. 8: 569-571.

[2] Behara A.M.P., Colling D.T., Gibson J.P. (1998a). Genetic diversity in endangered and common Canadian horse breeds. *Animal Genetics* 29 (Suppl. 1), str. 16.

[3] Bowling A.T., Eggleston-Stott M.L., Byrns G., Clark R.S., Dileanis S., Wictum E. (1997). Validation of microsatellite markers for routine horse parentage testing. *Animal Genetics* 28, str. 247–252.

[4] Canon J., Checa ML., Carlos C., Vega-Pla JL., Vallejo M., Dunner S. (2000). The genetic structure of Spanish Celtic horse breeds inferred from microsatellite data. *Animal Genetics,* 31, str. 39-48.

[5] Dieringer, D.; Schlötterer, C., 2002: Microsatellite analyser (MSA): a platform independent analysis tool for large microsatellite data sets. *Molecular Ecology Notes* 3 (1): 167-169

[6] Hamanova K., Majzlik I., Glasnak V., Schröffelowa D. (2001). Characterization and comparison of some horse breeds in Czech Republic as based on microsatellite markers polymorphism. *Animal Science Papers and Reports* 19 (3): 219-230.

[7] Jackowski M. (2002). Rozwój hodowli koni Huculskich w Polsce w okresie powojennym. Konferencja naukowa: *Nowe trendy w organizacji hodowli i rozrodu koni w Polsce*. Akademia Rolnicza w Krakowie, Kraków 6-7 Września 2002, str. 10 – 14.

[8] Kawasaki, E.S., 1990: Sample preparation from blood, cells and other fluids. In: Innis, M.A.; Gelfand, D.H.; Sninsky, J.J.; White, T.J. (ed.), *PCR Protocols: a Guide to Methods and Applications,* Acad. Press. New York, pp 146-152.

[9] Miller, M.P., 1997: Tools for population genetic analysis [TFPGA] 1.3: A Windows program for the analysis of allozyme and molecular population genetic data. Computer software distributed by author.

[10] Nei, M. (1978) Estimation of average heterozygosity and genetic distance from a number of individuals. *Genetics* 89: 538-590

[11] StockMarks® for Horses Equine Paternity PCR Typing Kit. Revision A, July 1996. *The Perkin-Elmer Corporation* (Applied Biosystems), Foster City, California (USA).

[12] Weir, B.S. and Cockerham, C.C. (1984) Estimating F-Statistics for the Analysis of Population Structure, *Evolution* 38(6): 1358-1370

[13] Ząbek T., Duniec M., Bugno M. (2003). Genetic relationships between Silesian, Thoroughbred and Oldenburg horses based on DNA microsatellite polymorphism. In *Annals of Animal Science*. Vol. 3, No. 2, 213-224.

In: Biotechnology, Agriculture and the Food Industry
Editor: G. E. Zaikov, pp. 21-25 ISBN 1-60021-040-6

Chapter 4

Analysis of Microsatellite Sequences Conservatism Between Family Bovidae and Cervidae

Agata Żyga, Anna Radko, Ewa Słota and Wojciech Włosiński*

Department of Immuno- and Cytogenetics, National Research Institute of Animal Production, ul. Krakowska 1, 32-083 Balice, Poland

* Polish Hunting Association, Cracow, Poland

Abstract

Conservatism of microsatellite sequences in Bovidae and Cervidae was analyzed using 11 microsatellite loci recommended by the International Society for Animal Genetics (ISAG) for cattle pedigree testing: TGLA227, BM2113, TGLA53, ETH10, SPS115, TGLA126, TGLA122, INRA23, ETH3, ETH225, BM1824

Based on the isolated genomic DNA, selected sequences were amplified in 6 samples of Cervidae by the polymerase chain reaction (PCR) using fluorescent labeled primers.The PCR products were separated by electrophoresis in a 4% polyacrylamide gel on a laser sequencer ABI PRISM 377.

The initial test indicate that cattle microsatellite markers may be useful for genetic description and phylogenetic analyses of the Cervidae.

Key words: *microsatellite sequences, Bovidae, Cervidae, genetic conservatism*

Introduction

Some authors try to distinguish a number of subspecies or ecotypes within the species of the *Cervidae* family (Janczewski at al, 1981). Studies at the molecular level, referring to the polymorphism of DNA, may give considerable information about the genetic differentiation of red deer (Cervus elaphus) and roe deer (Capreolus capreolus) and thus help to identify specific ecotypes (Niemczewski, 2002).

The greatest tools for these investigations are microsatellite sequences of DNA. Microsatellite sequences that are 1-5 nucleotide tandem repeats have found the widest application for animal genetic study to analyze genetic structure, variability and individual identification of animal.

Thanks to their high polymorphism and relatively simple identification, they became the largest class of genetic markers (Martin-Burriel et al., 1999; Maudet et al., 2002; Radko et al., 2002)

Comparative researches between the *Bovidae* and the *Cervidae* are increasingly important tools in searching for new genetic markers in the Cervidae with utilizing information available from a well-mapped species to species in which genetic mapping is not developed (Slate at al, 2002).

The aim of the present study was to analyze possibility of utilization of bovine 11 microsatellite markers to the family Cervidae.

Material and Methods

The investigation included six animals: four roe deer (Capreolus capreolus) and two red deer (Cervus elaphus)from forests localized in Southern Poland.

The blood samples were taken directly *post mortem* from animals shot during hunting events, into tubes containing EDTA, and stored at -20°C.

Genomic DNA was isolated using proteinase K, following the method of Kawasaki (1990).

Based on the genomic DNA conservatism between Bovidae and Cervidae at 11 loci was examined. Selected microsatellite sequences are recommended by the International Society for Animal Genetics (ISAG) for cattle pedigree testing: TGLA227, BM2113, TGLA53, ETH10, SPS115, TGLA126, TGLA122, INRA23, ETH3, ETH225, BM1824. The primers that amplify these microsatellites were contained in the "StockMarks for Cattle II" kit (Applied Biosystems). The PCR reaction was performed with the following reagents: 10x PCR Buffer, dNTPmix (a mixture of 4 deoxyribonucleotides at 1.25 mM each), AmpliTaq Gold DNA polymerase (5U/μL) and a mixture of 11 pairs of primers labeled with fluorescent dyes (2.5 pmol of each sequence).

The multiplex PCR reaction were performed in a Perkin Elmer GeneAmp System 9600 thermocycler with the following cycling parameters: 95°C for 15 min, followed by 31 cycles of 94°C for 45 s, 61°C for 45 s, 72°C for 60 s, and final extension at 72°C for 60 min.

The PCR products were separated by electrophoresis in a 4% polyacrylamide gel on a laser sequencer ABI PRISM 377. The results were analyzed with GeneScan 2.1 software and the identified alleles were sized with Genotyper 2.0 software.

RESULTS AND DISCUSSION

Among the 12 loci examined, 39 alleles was identified: at loci TGLA227 (2 alleles), BM2113 (4 alleles), TGLA53 (6 alleles) , ETH10 (3 alleles) , SPS115 (3 alleles) , TGLA126 (5 alleles), , TGLA122 (alleles 2) , INRA23 (3 alleles), ETH3 (6 alleles), ETH225 (2 alleles), BM1824 (3 alleles). Observed alleles at microsatellite loci within the group of examined animals in this study are shown in Table 1. The three loci: ETH10, TGLA122, and INRA23 were more difficult to amplification than the other.

Table 1. Alleles observed at microsatellite sequences loci within the group of animals examinated in this study

Marker	Alleles identified
TGLA227	63 69
BM2113	123 125 131 139
TGLA53	173 175 177 181 183 187
ETH10	217 219 221
SPS115	244 246 258
TGLA126	109 111 113 117 123
TGLA122	129 143
INRA23	204 206 214
ETH3	95 99 105 111 117
ETH225	137 149
BM1824	174 178 180

In the Table 2. we compare alleles identified in this study and results of Niemczewski et al (2002) at the five loci (TGLA53, TGLA126, TGLA122, ETH3 and ETH225).

The size range of alleles identified in this study are similar to those given by Niemczewski.

Our results are preliminary and should be continued but they suggest that chosen cattle microsatellite markers may be useful for the genetic description of the Cervidae.

Table 2. Alleles identified in this study compared to those given by Niemczewski *et al.*

Locus	Alleles identified in this study	Alleles after Niemczewski at al
TGLA53	173	163
	175	167
	177	169
	181	171
	183	173
	187	175
		177
		179
		183
		187
Size range (bp)	173-187	163-187
TGLA126	109	99
	111	111
	113	
	117	
	123	
Size range (bp)	109-123	99-111
TGLA122	129	139
	143	171
		175
Size range (bp)	129-143	139-175
ETH3	95	99
	99	101
	105	
	111	
	117	
Size range (bp)	95-117	99-101
ETH225	137	139
	139	
Size range (bp)	137-139	139

REFERENCES

[1] Z. Janczewski , *PWRiL* (1981), (In Polish).
[2] Martin-Burriel, E. Garcia-Muro, P. Zaragoza: *Anim Genet.*,30 (3), 177-182p (1999).
[3] C. Maudet, G. Luikart, P. Taberlet: *J. Anim. Sci.* 80 (4), 942-950p, (2002).
[4] Radko, M. Duniec, T. Ząbek, Janik A., M. Natonek, *Medycyna Wet.*, 58, 708-710p, (2002) (In Polish).
[5] E. S. Kawasaki: Academic Press, New York,146- 152 (1990).
[6] C. Niemczewski, J. Rudkowski, M. Żurkowski: *Animal Science Papers and Reports*, 20, 169-174p (2002).
[7] J. Slate, T. C. Van Stijn, R. M. Anderson, K. M. Mcewan, N. J. Maqbool, H.C.Mathias, M. J. Bixley, D. R. Stevens, A. J. Molenaar, J. E. Beever, S. M. Galloway, M. L. Tate: *Genetics* 160, 1587-1597p, (2002).

In: Biotechnology, Agriculture and the Food Industry
Editor: G. E. Zaikov, pp. 27-33 ISBN 1-60021-040-6

Chapter 5

PECULIARITIES OF MICROPROPAGATION OF SOME REPRESENTATIVES OF THE GENUS *HOSTA* TRATT. BY THE CULTURE OF OVARIES

A. Y. Nabieva[*] *and L. V. Buglova*
Central Siberian Botanical Garden

ABSTRACT

Among four types of explant source of Hosta tested, only ovaries from immature florets produced differentiated shoots and roots (rhizoids). Cold treatment and darkness were nessesary for gemmation this is ok numerous meristematic buds when revealed gistollogically. Regenerants appeared on different parts of rhizoids during a five month Period. Morphogenetic reaction and rate of shoot multiplication greatly depended on these genotypes.

Key words: *micropropagation, ovary culture, regeneration, 2,4-D, BA, rhizoids, bud formation, Hosta.*

ABBREVIATIONS

mMS or mN6 - modified MS or N6 medium;
BA –Benzylaminopurine;
NAA - Naphthaleneacetic acid;
2,4-D - 2,4-Dichlorophenoxyacetic acid;
2–ip - 2-isopentenyl adenine.

[*] Central Siberian Botanical Garden, Russia, 630090 Novosibirsk, Zolotodolinskaya str., 101; E-mail: root@botgard.nsc.su; nay@csbg.nsc.ru, fax: (3832) 30-19-86

Introduction

Micropropagation of hosta using shoot-tip cultures [1,2,3] is associated with contamination and low propagation speed. Researchers selecting immature inflorescences (scapes) as an explant source were a little more successful [4], whereas the others using floral buds and immature florets could enhance propagation of H. plantaginea and H. sieboldiana in vitro [5,6]. R.A. Hill et al. [7], D.J. Williams et al. [8] found that excised ovaries were an improved explant source over those mentioned above. Their results showed that ovary culture allowed to reach the highest frequency of shoot initiation and 20 - 30 shoots of H. sieboldiana could be obtained from a single ovary. The excised ovaries produced callus [7] and strict development of shoots or roots was not observed. Non callus development was indicated in the work of O.A. Churikova et al.[9] dealing with the micropropagation of H. sieboldiana.

The purpose of this study was to reveal morphogenetic reaction of different kinds of floral explants of the genus Hosta Tratt. in connection with:

- their age;
- genotypic differences;
- combination of macrosolts and/or phytohormones in culture medium.

For preservation of a Hosta genotype, especially of its hybrids, it was necessary to establish physiological status of explants and cultural components which could predetermine the developmental pathways in vitro. For the same reason it is important to avoid the callus stage during shoot proliferation. Histological study of tissues giving rise to regenerants is useful for clarification of their origin and peculiarities of morphogenesis. The age of the florets able to undergo the vegetative way in vitro is connected with the definite stage of microsporocytes development. The latter we tried to estimate according to the bud length of each Hosta genotype tested.

Materials and Methods

Explant materials were obtained from 4 Hosta species: *H. sieboldiana* Engl., *H. lancifolia* Engl., *H.fortunei var. aureo-marginata* (Hyl.) and *H.undulate var.mediovariegata* and 4 cultivars: *H. 'Krossa Regal', H. 'Cristmas Tree', H. 'Striptease', H. 'Sum and Substance'.* All plants were vegetatively propagated in the experimental plot of Central Siberian Botanical Garden in the period from June 7 to July 26. Four types of floral explants were put in the experiment: immature florets, excised ovaries, pedicel's 2 mm sections and stamens with anthers. Florets were ranged in two scales: 7-20 mm and 20-32 mm in length. Ovaries were excised from florets by cutting away perianths. Florets were preconditioned during 1 - 3 days at 4°C in the dark. In the first variant all types of explants were based on the MS medium [10], modified in compliance with [7]: 170 mgl^{-1} of $NaH_2PO_4{\cdot}H_2O$ and 360 mgl^{-1} of KH_2PO_4 were added to main salts at MS protocol and 0.4 mgl^{-1} of thiamine, 2.0 mgl^{-1} of glycine and 3% w/vol of sucrose were added to the organics. The modified N6 [11] medium - mN6, commonly used for culturing anthers in vitro, was selected as the second variant of initiation culture. Its organics contribution was the same as in mMS. Both media were solidified with 0,6% Difco Bacto agar and the pH was adjusted to 5.7 prior to autoclaving (20 min at 121°C).

Florets with the pedicels were surface sterilized by immersion in 70 % v/v ethanol for 1 min followed by 0,2% mercuric chloride with a few drops of Tween 80 for 8 min and rinsed three times in sterile deionised water. At the initiation stage the explants were maintained (for a week in the dark at 26-28°C and then under fluorescent lights at 16-h light/8-h dark photoperiod at 25±1°C) on mMS and mN6 media with BA, 2-ip, NAA and 2,4-D in different combinations, as well as on mMS without phytohormones. Morphogenetic reaction of ovary explants was tested on two variants of initiation media, where 2,4-D or BA and 2-ip or NAA were added. Thus mMS was combined with BA and NAA and mN6 – with 2-ip and 2,4-D. The following auxin/cytokinin ratios - 1/2; 1/1; 2/1; 4/1; 5/1- were tested. Values of cytokinins were from 0,2 to 0,4 mg/l and those of auxins – from 0,2 to 1,0 mg/l. Each proliferating culture was left onto the initiation medium until regeneration or transferred after 1-1,5 month on the fresh one. Differentiation medium constituted the mMS medium but the ratio of the auxins/cytokinins was reversed. Histological study of tissues giving rise to regenerants was carried out by standard methods [12]. Meiocytes were stained by acid carmine. The tissues were stained by methyl green – pyronin to find RNA or DNA in the cells. Samples were viewed through a Zeiss microscope "Axiolab", microphotographs were made by "Axioscope 2+" equipped with "Axiocam HRc".

RESULTS AND DISCUSSION

The results showed that regardless a genotype among 4 types of floral explants only ovaries could produce differentiated shoots. Species morphogenetic reaction of ovaries differed in the following way: ovaries of the first 2 species- *H. sieboldiana* Engl. and *H. lancifolia* Engl. turned brown on all media tested. In two other species- *H.fortunei var. aureo-marginata* (Hyl.) (Fig.1) and *H.undulate var.mediovariegata* the morphogenetic reaction of explants during cultivation was different: after 2-3 months on initiation medium the first one developed only rhizome-like structures, and later after culturing on differentiation medium it emerged shoots on the base of the ovaries. The second species demonstrated strict development of shoots on all parts of rhizoids after cultivation for 3-4 months in the same conditions.

Compared with species, Hosta cultivars appeared to be more capable to organogenesis in vitro: in 3 of 4 cultivars (*H. 'Krossa Regal', H. 'Cristmas Tree'*(Fig.2), *H. 'Striptease', H. 'Sum and Substance'*) morphogenesis was similar to gemmorhizogenesis type [14] with reverse succession of events. The last cultivar developed secondary explants resembling altered carpels. The same way of ovary development was mentioned at [7] dealing with *H. sieboldiana.* In this work the secondary structure appearing on ovary explants considered to be parthenocarpic fruit.

Two media - mMS and mN6 were tested in the experiment by morphogenetic reaction of explants: their response to callus production, shoot and/or rhizoids initiation.

According to nutritional composition of mN6 and mMS no substational differences were found in the ovary development of all Hosta genotypes tested.

Fig.1. Shoot and root regeneration from the ovary of *H.fortunei var. aureo-marginata* Hyl

Fig.2. Shoot and root regeneration from the ovary of *H. 'Cristmas Tree'*

Thus, while nutritional composition does not play the main role in triggering the program of proliferation, prevailed quantity of auxins, especially 2,4-D over BA or 2-ip, leads dramatically to development of rhizome-like structures, and only after this- shoots, if the level of cytokinines increases in differentiation medium. The optimum level of auxins (NAA and 2,4-D) is 0,8mg/l, whereas the level of cytokinines must be much lower: 0,1-0,2mg/l in initiation medium. In differentiation media the ratio of the auxins/cytokinins was reversed and values of BA, 2-ip reached 1 mg/l.

In order to compare size of florets with definite stage of microsporocyte development, we picked out several morphometric parameters of the florets and inflorescences: 1) size of the florets; 2) presence of the scape; 3) coloration of the florets – we ranged florets in coloured and noncoloured (green) ones. In all cases when floret size was minimal and two other parameters had no positive meaning – regeneration had not occurred. If the ovaries were excised from the well-grown florets which were gaily coloured – ovaries on nonhormonal mMS medium performed the features of floral morphogenesis and parthenocarpic fruits were formed. Only ovaries from noncoloured florets of middle size proliferated microshoots vegetatively. Microsporogenesis was studied only in that group of florets. Meiocytes were found to be at the stage of two nuclei. Pollen cell wall was not formed yet.

Morpho-anatomic study of regeneration showed that rhizoids, appearing the first, had a structure typical of roots [13]. Their surface was covered by rhizodermis and the parenchyma was found lower (Fig 3a, 3b). Central cylinder consisted of xylem vessels and elongated cells with large nuclei which had the ability for division (Fig 3b). We found that these cells were morphologically similar to the pericycle ones and secondary meristem. Hosta xylem was represented by spiral vessels. Radical hairs were formed from the cells of rhisodermis (Fig. 4, Fig. 5), which remained viable for a long time, and as a result of it - the extensive area of absorption arised. Buds were formed from meristematic cells of central cylinder (Fig. 4). At the first stage small tubercle was generated from cells of primary meristem, dividing in the fast and random way. In those cells an active synthesis of DNA was observed, but RNA was absent. When the scale primordium was initiated, its cells rapidly lost the ability to divide.

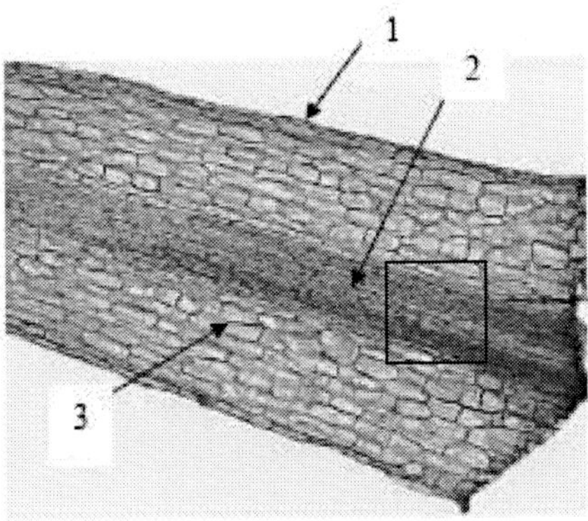

Fig.3. Rhizoid: 1.Rhisodermis 2.Central cylinder 3. Parenchyma

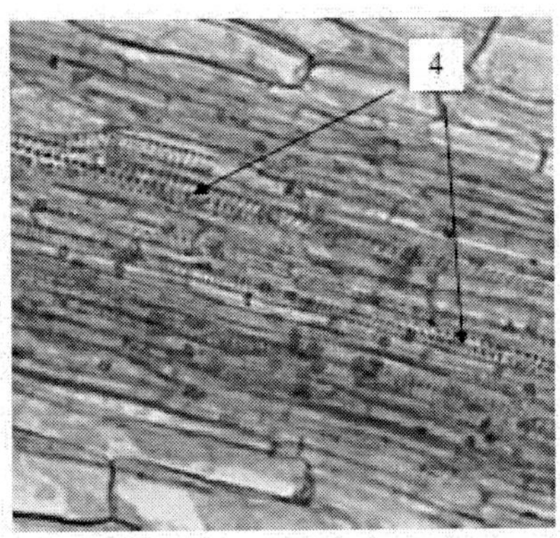

Fig.3b. Central cylinder: spiral vessel of xylem

Fig. 4.

Fig.4. The beginning of a bud formation

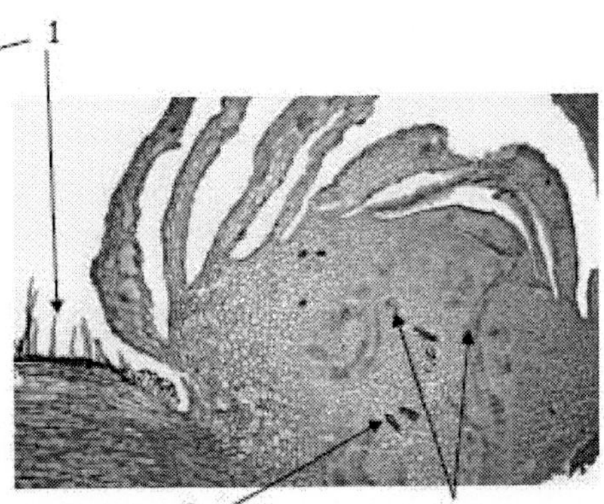

Fig. 5.

Fig.5. The bud of *H. 'Cristmas Tree'*, proliferated *in vitro* .1.Radical hairs 2.Storage cells 3. Meristem

As stated above [9], histological differentiation begins at the early stages of bud development. Meristematic cells create local areas: transverse and longitudinal ones around xylem vessels (Fig. 5). When the division rate decreases, meristem acquires the features of the secondary structure.

Conclusions

Among all the types of explants only ovaries represented proliferation and shoot regeneration occurring in two Hosta species and three hybrids. The morphogenesis belongs to the gemmorhizogenesis type and callusogenesis was not observed during ovary proliferation. Independently of media composition the prevailed content of auxin (2,4-D) over cytokinine (BA) promoted the formation of rhizome-like structures- (rhizoids). Shoot proliferation by generation of meristematic cells of central cylinder was mainly noted in different parts of rhizoids. These cells remained capable to divide for a long time. Further experiments allowed to develop a technology of Hosta micropropagation by ovary culture.

References

[1] M. Lubomski: *Acta Horticulturas J.*_251, 223-228 (1989). Shoot multiplication and rooting of Hosta sieboldiana cv. Gold Standard using cultured shoot tips.

[2] M. Papachatzi P., A. Hammer, P. M. Hasegawa: *J. Amer. Soc. Hort. Sci.* 106(2), 232-236 (1981). In vitro propagation of Hosta decorata 'Thomas Hogg' using Cultured Shoot Tips.

[3] K.Y. Paek, S.H. Ma: *Acta Horticulturas J.* 440, 576-581 (1996). In vitro propagation of hosta using cultured shoot tips and somaclonal variability of regenerants.

[4] M. R. Zilis, D. Zwagerman: *HortScience J.* 14(3), 456 (1980). Clonal propagation of hosta by scape section in vitro. (Abstract)

[5] M. Papachatzi P., A. Hammer, P. M. Hasegawa: *HortScience J.* 15(4), 506-507(1980). In vitro propagation of Hosta plantaginea.

[6] M. M. Meyer: *HortScience J.* 15(6),737-738 (1980). In vitro propagation of Hosta sieboldiana.

[7] R.A. Hill, G.A. Tuskan, A.A. Boe: Plant Cell, *Tissue and Organ Culture J.* 17, 71-75 (1989). In vitro propagation of *Hosta sieboldiana* using excised ovaries from immature florets.

[8] D.J. Williams, K.H. Al-Juboory, R.M. Skirvin: In *vitro Plant J.* 34 (4) 289 (1998). Adventitious shoot regeneration from ovaries of Hosta Golden Scepter.

[9] O.A. Churikova, O.I. Molkanova. Materials III Inter. Sci. Confer. *"Biodiversity and plant introduction"*. St. Petersburg, 2003, 288-289. Some regularities of daylily and hosta morphogenesis *in vitro.*

[10] T. Murashige, F. Skoog: *Physiol Plant J.* 15, 473-497 (1962). A revised medium for rapid growth and bioassays with tobacco tissue cultures.

[11] C.-C. Chuang., T.-W. Quyang. *//Proceed. Symp. on Plant Tissue Culture. Peking.*, Science Press, 1978, 51-56. A set of potato media for wheat anther culture.

[12] R.P. Barykina, T.D. Veselova, A.G. Devyatov, Kh.Kh. Dzhalilova, G.M. Ilyina, N.V. Chubatova. *A Handbook to botanical microtechnics*. Main metho*ds*. MSU, Moscow, 2004, 312 p. (in Russian).

[13] K. Esau. *Plant anatomy*. Moscow, 1969, 660 p. (in Russian).

[14] T. Batygina (Ed.) Embryology of flowering plants. *Terminology and concepts.* vol.2, World and family- 95, St. Petersburg, 1997, 1164 p. (in Russian).

In: Biotechnology, Agriculture and the Food Industry
Editor: G. E. Zaikov, pp. 35-39
ISBN 1-60021-040-6

Chapter 6

DNA MARKERS FOR INTROGRESSING RESISTANCE TO POTATO LATE BLIGHT

M. P. Beketova, V. A. Biryukova, N. S. Velishaeva, P. E. Drobyazina, V. S. Zaitsev, V. V. Martynov, E. E. Khavkin*, I. A. Shilov and G. A. Yakovleva*

*Institute of Agricultural Biotechnology,
Russian Academy of Agricultural Sciences, Moscow, 127550 Russia
**Institute of Potato Industry, Samokhvalovichi. Belarus

ABSTRACT

To follow the introgression of genetic material conferring late blight resistance from *Solanum bulbocastanum* and *S. demissum* into potato hybrids, we used PCR technologies based on the amplification of the *RB* and *R1* gene sequences and the non-translated sequences (microsatellites and dispersed repeats) that flank the corresponding genes.

Key words: *Solanum bulbocastanum - Solanum demissum - Solanum tuberosum - Phytophthora infestans - R genes - microsatellites - dispersed repeats*

Late blight caused by the oomycete *Phytophthora infestans* is the most devastating disease of potato (*Solanum tuberosum* ssp. *tuberosum*). For several decades breeding potato cultivars resistant to late blight relied on crossing potato to its wild relatives, such as *S. demissum* that comprises the race-specific resistance (R) genes. According to the current opinion, the R-dependent resistance of potato cultivars to late blight was overcome due to pathogen evolution; however, such opinion is based on the phenotypic evidence rather than the molecular studies of host vs. pathogen genes. Recent cloning the *R1* [1] and *R3a* (Huang, in preparation) genes from *S. demissum* paved the road to such verification.

* Institute of Agricultural Biotechnology, Russian Academy of Agricultural Sciences, Timiryazevskaya ul., 42, Moscow, 127550 Russia; fax: 7 (095) 977-0947; e-mail: emil@iab.ac.ru

In their search for new sources of durable late blight resistance, potato breeders have concentrated on other *Solanum* species, such as *S. bulbocastanum,* with its *RB* gene conferring complete resistance to *Phytophthora* [2]. To transfer *RB* to potato, somatic hybrids of *S. tuberosum* and *S. bulbocastanum* are obtained by protoplast fusion. Two such hybrids were produced in the Institute of Potato Industry (Samokhvalovichi. Belarus) [3].

The molecular studies *R1*, *R3a* and *RB* and their genomic environment in *S. demissum* and *S. bulbocastanum* [1, 4-6] made it possible to detect these genes in potato genotypes obtained by introgression of *S. demissum* and *S. bulbocastanum* germplasm and to weigh up the phenotypic evidence on late blight resistance obtained under laboratory and field conditions directly against the presence, dosage and expression of these genes.

The development of DNA markers of late blight resistance based either immediately on the R gene fragments or on the genomic sequences that border the R genes would provide a reliable instrument for monitoring introgression of the genes of late blight resistance with the germplasm of wild *Solanum* species, e.g., in the segregating hybrid progenies at the seedling phase.

To monitor *R1,* we used the PCR technology worked out in the Gebhardt lab [1, 6] and verified it against race-specific differentials (Fig. 1) and also with such potato hybrids and cultivars wherein the presence *R1* was attested by pedigrees and by the tests with specific inocula of *Ph. infestans* (Beketova *et al.*, in preparation). By cloning the 1400-bp *R1* fragment from the potato hybrid 78563-76 derived from the demissoid potato cv. Apta, we confirmed the identity of this marker to the prototype *S. demissum* sequence (AF447489). The primers for PCR amplification of the *RB* gene were designed on the basis of the *RGA2* sequence (AY336128, [4]) and verified by comparing several accessions of *S. tuberosum* and *S. bulbocastanum*. The markers for *R1* and *RB* were used to monitor these resistance genes in two somatic *S. bulbocastanum* (+) *S. tuberosum* hybrids and their seed progenies (Fig. 2).

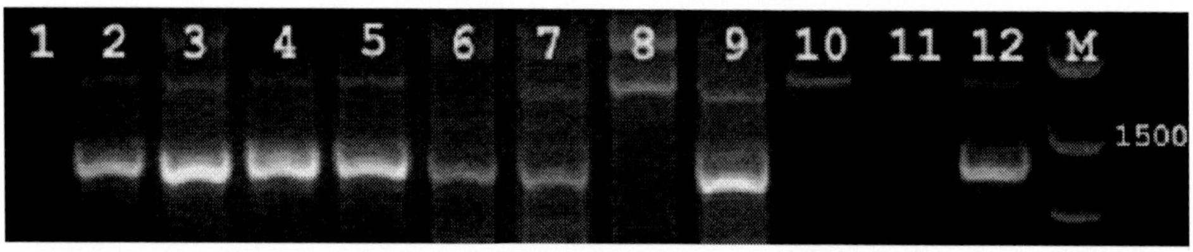

Fig. 1. Verification of the PCR test for the *R1* gene for potato late blight resistance 1 – control; 2-11 - CIP differentials: 2 – 801039 (R1); 3 – 801049 (R1); 4 – 801050 (R1); 5 – 801051 (R1); 6 – 801055 (R1); 7 – 801056 (R1); 8 – 801057 (R1); 9 – 801059 (R1); 10 – 801038 (r); 11 – 801045 (R7); 12 – potato cv. Temp; M – Marker 100 bp DNA Ladder Plus.

To develop DNA markers from non-translated genome sequences that flank the *R1* gene, we put side by side the polymorphic spectra of microsatellites and dispersed repeats in the demissoid potato forms and those accessions of *S. demissum* wherein the *R1* gene was registered by the PCR test. Similar approach was used to compare the parental forms of *S. bulbocastanum* and *S. tuberosum* to somatic *S. bulbocastanum* (+) *S. tuberosum* hybrids and their seed progenies. The primers for the amplification of microsatellite and dispersed repetitive sequences are described in the captions to Figs. 3 and 4.

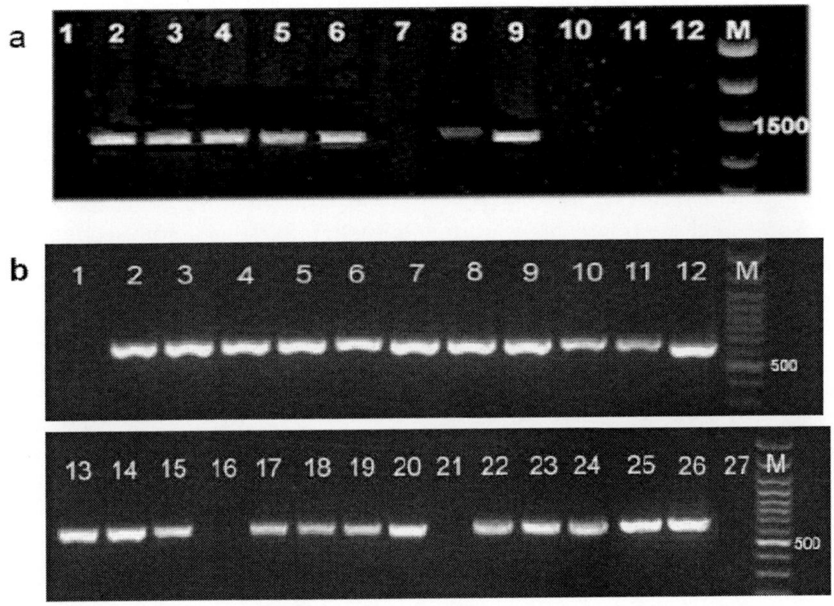

Fig. 2. Introgression of the R genes into the somatic hybrids *S. bulbocastanum* (+) *S. tuberosum* (a) *R1*: 1 – control; 2 – cv. Temp; 3-5 – somatic hybrids (3 – Sb6-7; 4 – Sb3-2; 5 – Sb7); 6 – hybrid 78563-76 (*S. tuberosum* parent); 7 – *S. bulbocastanum* parent; 8 – Sb5-2 (somatic hybrid 78563-76 + *S. bulbocastanum*); 9 – Sc5-19 (seed progeny); 10-11 - *S. tuberosum* pollinators (10 – hybrid 88.34/14; 11 – hybrid 88.16/20); 12 – cv. Belorusskii 3; M – Marker 100 bp DNA Ladder Plus. (b) *RB*: 1 – control; 2-8 – *S. bulbocastanum* (Sb45-3; Sb45-4; Sb45-6; Sb45-7; Sb45-9; Sb45-11; *S. bulbocastanum* parent); 9-15 – somatic hybrids (Sb1-1; Sb3-2; Sb6-7; Sb7; Sb11-1; Sb13-1; Sb5-2); 16 – hybrid 78563-76 (*S. tuberosum* parent); 17-18 – seed progeny Sb5-2 (Sc5-19; Sc5-22); 19-20 – free pollination of Sb5-2; 21 – pollinator hybrid 88.34/14; 22-26 – somatic hybrids (L4-11; L4-12; L4-16; L4-18; L22); 27 – cv. Lasunok (*S. tuberosum* parent); M – Marker 100 bp DNA Ladder. Plus. Primers: GGCAAGCTCAGAATCAAATTATCCACCCCAACTTTTAAAT and GATGCCCAACACCTTTCAGTCCTCCCAATACTTG.

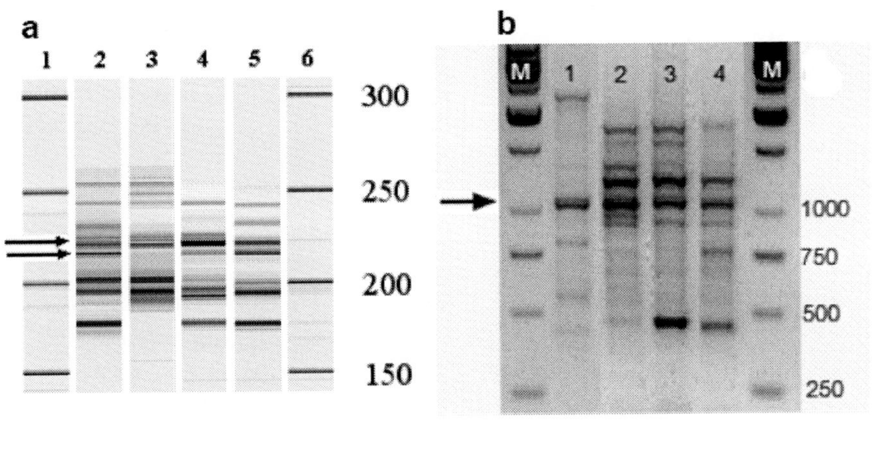

Fig. 3. Putative flanking markers of *S. demissum* germplasm (shown with arrows) (a) Microsatellites, primer STIIKA [7]. 1, 6 - Marker 50-500; 2-4 - potato cvs Effect, Golubizna, and Skoroplodnyi; 5 - *S. demissum*. (b) Dispersed repeats, primer PAWS5 [8]. M - Marker 100 bp DNA Ladder Plus; 1 - *S. demissum*; 2-4 - potato cvs Effect, Golubizna, and Skoroplodnyi.

The PCR analyses provided sound proof of *R1* and *RB* introgression into the somatic hybrids from the corresponding parental forms (Fig. 2). We also found characteristic microsatellites and dispersed repeats of the *S. demissum* and *S. bulbocastanum* genomes apparently transferred to the sexual and somatic hybrids and maintained in their progenies when selected for their resistance to late blight (Figs. 3 and 4). These DNA fragments were cloned, and their sequences were compared to the NCBI and TIGR genomic and EST databases. Some of these fragments were found to align to the genomic sequences mapped in the vicinity of the *R1* and several unidentified genes for late blight resistance (Biryukova *et al.*, in preparation).

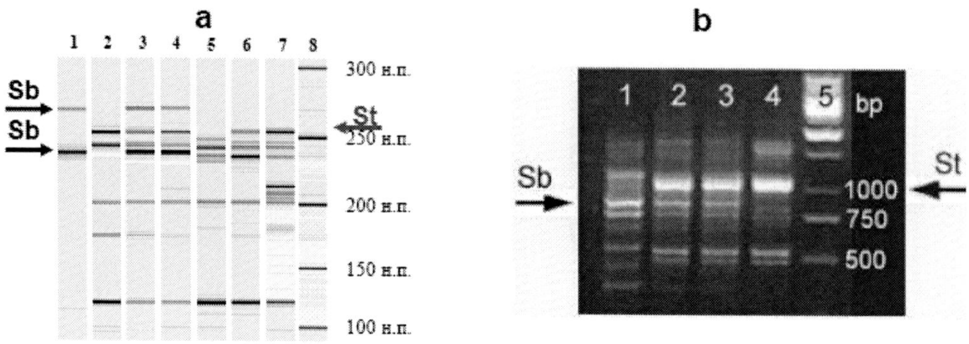

Fig. 4. Putative flanking markers of *S. bulbocastanum* germplasm (shown with arrows) (a) Microsatellites, primers STS 1/2 [7]. 1 - *S. bulbocastanum*; 2 - *S. tuberosum*, cv. Lasunok; 3, 4 - L 4-12 and L 4-11 somatic hybrids *S. bulbocastanum* (+) *S. tuberosum*; 5 - hybrid 78563-76 (*S. tuberosum* parent); 6, 7 - *S. tuberosum* pollinators 88.16/20 and 88.34/14; 8 - molecular weight marker 50-500. (b) Dispersed repeats, primer PAWS6 [8]. 1 - *S. bulbocastanum*; 2, 3 - L4-11 and L4-12 hybrids *S. bulbocastanum* (+) *S. tuberosum*; 3 - cv. Lasunok (*S. tuberosum* parent); 4 - marker GeneRuler 1 kb DNA Ladder.

References

[1] Ballvora, A., Ercolano, M.R., Weiss, J. *et al.* The R1 gene for potato resistance to late blight (*Phytophthora infestans*) belongs to the leucine zipper/NBS/LRR class of plant resistance genes. *Plant J.*, 2002, 30: 361-371.

[2] Grünwald, N.J., Hinojosa, M.A.C., Covarrubias, O.R. et al. Potato Cultivars from the Mexican National Program: Sources and Durability of Resistance against Late Blight. *Phytopathology*, 2002, 92: 688-693.

[3] Yakovleva, G.A., Semanyuk, T.V., Monarkhovich, S.V., Podobed, N.I. Late blight resistant somatic hybrids of *Solanum tuberosum* (48, 24) with *Solanum bulbocastanum.* GILB'02 Conference Late Blight: Managing the global threat, *Poster Abstracts,* Hamburg, 2002.

[4] Song J., Bradeen, J.M., Naess, S. K. *et al.* Gene RB cloned from *Solanum bulbocastanum* confers broad spectrum resistance to potato late blight. *Proc. Nat. Acad. Sci.*, 2003, 100: 9128-9133.

[5] Van Der Vossen Eag, Sikkema, A., Hekkert B.T.L. *et al.* An ancient *R* gene from the wild potato species *Solanum bulbocastanum* confers broad-spectrum resistance to *Phytophthora infestans* in cultivated potato and tomato. *Plant J.*, 2003, 36: 867-882.

[6] Gebhardt, C., Ballvora, A., Walkemeier, B. *et al.* Assessing genetic potential in germplasm collections of crop plants by marker-trait association: a case study for potatoes with quantitative variation of resistance to late blight and maturity type. *Mol. Breed.*, 2004, 13: 93-102.

[7] Mcgregor, C.E., Lambert, C.A., Greyling, M.M. e.a. A comparative assessment of DNA fingerprinting techniques (RAPD, ISSR, AFLP and SSR) in tetraploid potato *(Solanum tuberosum* L.) germplasm. *Euphytica,* 2000, 113: 135–144.

[8] Rogowsky, P.M., Shepherd, K.W., Langridge P. Polymerase chain reaction based mapping of rye involving repeated DNA sequences. *Genome,* 1992, 35: 621-626.

In: Biotechnology, Agriculture and the Food Industry
Editor: G. E. Zaikov, pp. 41-45
ISBN 1-60021-040-6

Chapter 7

THE IMPROVED SYNTHESIS OF EPIBRASSINOLIDE AND ITS APPLICATION IN THE AGRICULTURE

A. V. Cheprakov*, M. A. Filatov*, N. V. Lukashev*, I. P. Beletskaya* and N .N. Malevannaya**

*Moscow State Lomonosov University,
Department of Chemistry 119992, Moscow
**NNPP «NEST-M», Moscow

ABSTRACT

A new effective synthetic approach to phytohormone 24-epibrassinolide has been described. The method exploits the potential of aqueous microheterogeneous solubilizing media for the realization of high yield reduction of the key intermediate, iso-ergosterone, and thus enables to produce the target phytohormone of high purity, required for commercial implementation. Epibrassinolide is the active ingredient of antistress adaptogenic plant growth regulator EPIN-Extra. The application of EPIN-Extra in various agricultural areas is described.

Keywords: *epibrassinolide, microemulsion, phytohormone, plant growth regulator, adaptogen, agriculture, horticulture, honey bee keeping, fish breeding*

Epibrassinolide belongs to brassinosteroids, a recently discovered class of phytohormones. BS in very low concentrations (10^{-12}–10^{-7} M) are known to facilitate germination of seeds, promote growth and development of plants, increase tolerance to stress factors, thus having positive effect on crops[1; 2].

As soon as brassinosteroids are widely occurring in Nature, these compounds at naturally compatible concentrations should be regarded as environmentally safe compounds due to by default participation in food chains. Therefore, brassinosteroids should be one of primary targets of research for developing a new generation of biorational, environmentally friendly

* Moscow State Lomonosov University, Department of Chemistry 119992, Leninskie Gory, Moscow, e-mail: avchep@elorg.chem.msu.ru

highly effective crop protection agents. For Russia, with its highly variable harsh climate and exhausted ecosystems brassinosteroids are particularly valuable. The investigation and implementation of brassinosteroids and other biorational agents open possibilities for the development of the sustainable agriculture [3], free from hazardous and environmentally unwelcome chemical agents.

The application of brassinosteroids at extremely low rates (ca. 1-2 mg/h) should allow for

i. essential increase of tolerance of agricultural plants to unfavorable conditions including frost, temperature shocks, drought, excess of moisture, low insolation, pollution of soil and air, high salinity of soil, etc., thus resulting in essential increase of crops under far from optimal conditions.
ii. increase of resistance towards a wide range of fungal and microbial plant deceases
iii. decrease of application rates of chemical pesticides
iv. improvement of crops quality

Epibrassinolide has been tested in various agricultural areas for different types of crops. The treatment of seeds by epibrassinolide has been proven to result in better rates of germination, enhanced propagation, faster biosynthesis of bioactive compounds, and better resistance to pathogens. The treatment of potato tubers before planting along with a single spraying of growing plants by epibrassinolide solution helps growing healthy crops free of fungal deceases, which results in better market value and decreased losses during storage and transportation.

In vineyards, spraying by epibrassinolide solution has been found to result in the promotion of growth of plants, increase of average weight of grapes, net increase of crops by ca. 50%. Unlike common growth stimulators epibrassinolide also positively influenced such important parameter of grape vines as the content of sugars, effecting more than 1% increase.

Particularly interesting is the application of epibrassinolide in floriculture, city landscaping, professional and amateur decorative gardening. The environment in big cities like Moscow is particularly damaging for plants due to eroded ecosystems, high pollution of air and soil. The application of epibrassinolide should allow for the restoration of damaged plants, increased tolerance towards stresses. Regular treatment of decorative plants, which often belong to other climatic zones, results in better degree of adaptation to new conditions of growing, helps to increase the survival rate after winter.

Besides crop protection, epibrassinolide finds interesting applications in other fields of agriculture. It should be noted that brassinosteroids are ubiquitous compounds in plants, thus being normally consumed by all herbivorous animals, including insects, fish, and mammals. The application of epibrassinolide as food additive helps to increase resistance of various animals towards toxic pollutants in the environment.

This effect has already found practical application in fish breeding farms near Volga and Caspian sea. The addition of epibrassinolide to technological water used in breeding cisterns markedly (by over 2 times) increases life expectancy of embryos, larvae and spawn of the sturgeon.

The other proven effect of epibrassinolide is cryoprotection, that is the increase of tolerance of semen towards freeze shock, which is of high importance for the storage and dissemination of genetic material of valuable breeds of fish, horses, etc. through banks of frozen semen.

In the bee-keeping the application of epibrassinolide helps to compensate for deficiencies of natural feed-stocks in winter/spring season. In order to maintain high productivity of bee families, the deficiency of natural plant hormones common in pollen of honey plants can be compensated for by the addition of epibrassinolide to sugar syrup used for feeding bees off-season. This treatment has been found to result in 70% boost of bee queen egg-laying ability, as well as in the prolongation of life expectancy of working bees. The effect of epibrassinolide is particularly notable in the areas with short summer, where the bees are not adequately supplied with the natural pollen hormones.

Improved synthesis of epibrassinolide. The published synthesis of epibrassinolide is a multistep procedure, starting from readily available yiest ergosterine (1)[4; 5]. One of key stages of this synthesis is the selective transformation of iso-ergosterone (3) into (22E, 24R)-3α,5-cyclo-5α-ergosta-22-en-6-one (4).

Scheme

In spite of long synthetic efforts devoted to optimization of this reaction, it still presents the main obstacle in the synthetic scheme due to very poor reproducibility, selectivity and productivity (cf. e.g. [6-8]). Practical realization of the synthetic approach to epibrassinolide depends on improvement of this stage. Using the technology of aqueous solubilizing media earlier developed by us[9; 10], we have elaborated an improved method of synthesis, which affords better reproducibility, essential enhancement of technological parameters, overall to make the process economically feasible.

After a study of a wide scope of reducing agents described in the literature for conjugate reduction of enones, we have chosen sodium dithionite as the most promising reagent. Sodium dithionite is a very cheap readily available reducing agent capable of selective reduction of a number of important functional groups. However, a considerable preparative potential of this reducing agent remained largely unexplored because of its insolubility in non-aqueous media, which effectively interferes with the application of this reagent for water-insoluble organic substrates. The use of aqueous solvents, e.g. DMF allows to perform reductions of only relatively small compounds which are able to at least partially dissolve in aqueous solvents with high water content. More hydrophobic compounds, such as steroids, give very poor results under such conditions due to extremely low solubility in solvents containing over 10 vol% water required for sodium dithionite. The application of phase-transfer techniques for dithionite reductions is ineffective. Highly hydrophilic dithionite anion can be transported to aqueous phase only in the presence of huge excess of very effective phase-transfer agents, such as *Aliquat 336* or *Adogen 464* [7; 11-13].

The use of the most thoroughly studied representatives of solubilizing systems, the microemulsions, as media for reactions is a so far poorly investigated area. There are still practically no investigations showing the potential of microemulsions or related solubilizing media for preparative-scale organic synthesis.

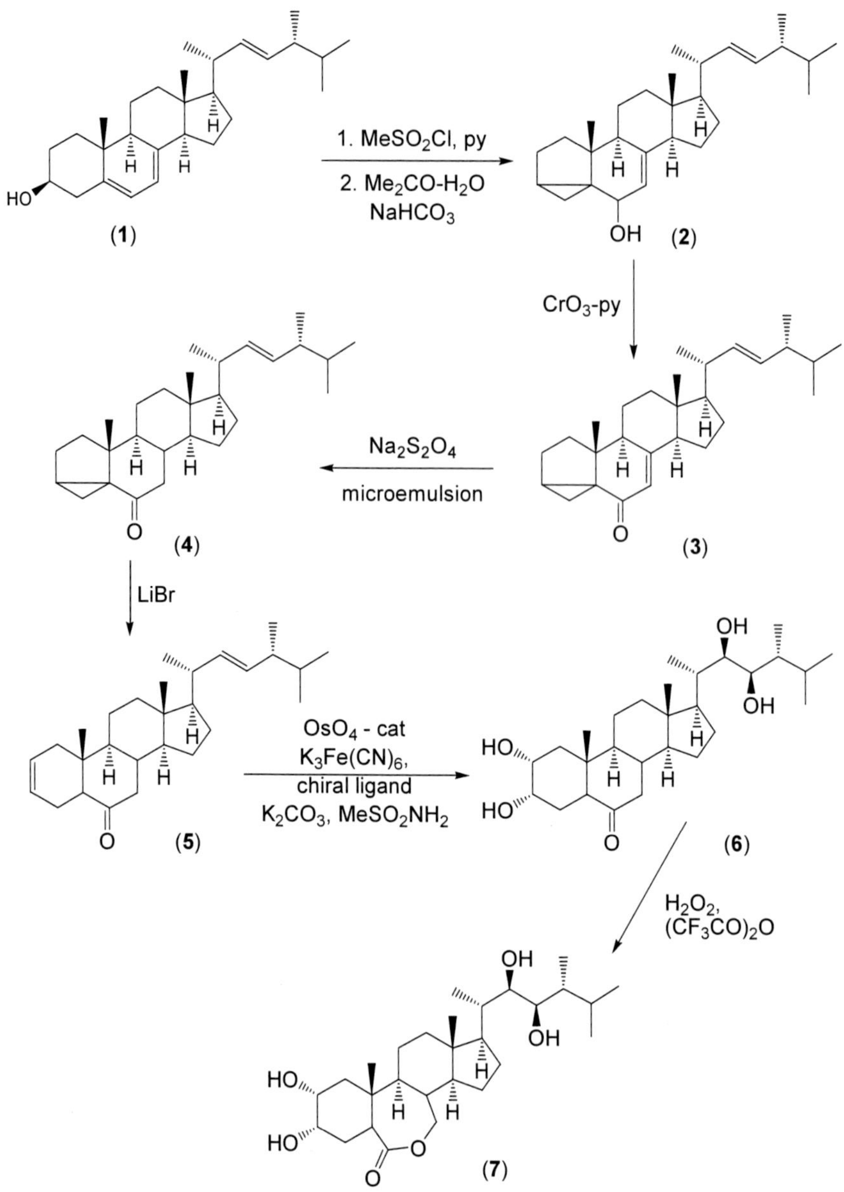

The solubilizing media are formed by mixing water, surfactant (ionic or non-ionic single-tail tenside), co-surfactant (usually a short chain alcohol), and electrolite in order to achieve the balance between hydrophilic and lipophilic properties of the system. The resulting systems are thermodynamically stable liquids capable of solubilization of high amounts of hydrophobic organic compounds. One of the most interesting features of solubilizing media is the ability to serve as media for reactions between the reagents of incompatible solubility profiles, e.g. between a highly hydrophobic organic substrate and an ionic water-soluble reagent. In our case this advantage is critical. Dithionite anion is well known to be a so highly hydrophilic species, that a conventional phase-transfer using long chain quaternary ammonium salts is ineffective. Meanwhile, we have shown than in a properly formulated microheterogeneous solubilizing media composed of single tail surfactant, n-butanol, water, and electrolyte ($NaHCO_3$) the reduction of steroid enone (3) takes place readily to give the expected product in high yields and selectivity. The outcome of reaction depends on the nature of solubilizing media. We have established that the best yields exceeding 80% are obtained in a single phase bicontinuous system corresponding to perfectly balanced hydrophile-lipophile interactions. The method has been proven to be reproducible, fully useful for large scale preparations, and thus for the first time opened up a possibility to realize practical synthesis of epibrassinolide of high quality at marketable cost.

The developed synthesis has been implemented for the preparation of active ingredient of commercial agent EpinExtra, developed and marketed by NNPP NEST-M in Moscow since 2002.

REFERENCES

[1] S. Hayat, A. Ahmad: *Brassinosteroids: Bioactivity and Crop Productivity*. Kluwer, N.Y., 2003.

[2] V. A. Khripach, V. N. Zhabinskii, A. De Groot: *Brassinosteroids--A New Class of Plant Hormones*. Academic Press, N.Y., 1999.

[3] W. M. Nelson: Green chemistry and the path to sustainable agriculture. ACS symposium series. 887, 7 (2004).

[4] T. C. Mcmorris, P. A. Patil: *J. Org. Chem*., 58(8), 2338 (1993).

[5] M. J. Thompson, N. Mandava, J. L. Flippenanderson, J. F. Worley, S. R. Dutky, W. E. Robbins, W. Lusby: *J. Org. Chem*., 44(26), 5002 (1979).

[6] C. Brosa, R. Puig, X. Comas, C. Fernandez: *Steroids,* 61(9), 540 (1996).

[7] C. Brosa, C. Rodriguez-Santamarta: *Tetrahedron*, 55(6), 1793 (1999).

[8] C. Brosa, C. Rodriguez-Santamarta, J. F. Pilard, J. Simonet: *Phys. Chem. Chem. Phys*., 3(13), 2655 (2001).

[9] I. P. Beletskaya, A. V. Cheprakov. (2002). Aqueous Palladium Catalysis. In N. E.I. (Ed.), *Handbook of Organopalladium Chemistry*, E.I. Negishi (Ed.) Wiley Int., N.Y., 2002.

[10] A. V. Cheprakov, N. V. Ponomareva, I. P. Beletskaya: *J. Organomet. Ch*em., 486(1-2), 297 (1995).

[11] F. Camps, J. Coll, J. Guitart: *Tetrahedron,* 42(16), 4603 (1986).

[12] F. Camps, J. Coll, J. Guitart: *Tetrahedron*, 43(10), 2329 (1987).

[13] F. Camps, J. Coll, J. Guitart: *J. Chem. Res*. (S), (1), 38 (1990).

In: Biotechnology, Agriculture and the Food Industry
Editor: G. E. Zaikov, pp. 47-54
ISBN 1-60021-040-6

Chapter 8

ISOLATION AND STUDY OF THERMOPHILIC MICROMYCETES-PRODUCERS OF ENZYMES AND PROTEIN

J. Tashpulatov*, B. G. Baybaev, T. S. Shulman, T. Abdullae and J. J. Tashpulatov
Institute of Microbiology, Tashkent, Uzbekistan

ABSTRACT

From water samples taken from thermal springs of Uzbekistan 20, strains of thermophilic micromycetes were isolated. They actively grow on cellulose and lignin containing agricultural wastes at high temperature (40-55°C). It was established that these strains grow on medium with addition of corn steep, wort and autolyzed yeasts. Studied thermophilic micromycetes assimilated glucose, sucrose, maltose, starch; many of them used cellulose. From nitrogen sources they well assimilated $NaNO_3$ both alone and in mixture with other organic sources in nutrient medium. The comparative study of fractional and amino acid composition in a the protein of *Thermomyces lanuginosus* strains this is ok conducted. It was established that *Th.lanuginosis*-15z a is high protein containing strain that produces 6.75 mg/ml of protein and full complex of contiguous and essential amino acids.

Keywords: *thermophilic and thermotolerant micromycetes, cellulase activity, protein production, amino acids*

Currently, thermophilic micromycetes are actively studied as producers of different physiologically active compounds. Unique peculiarities of thermophilic fungi – fast growth and ability to carry out different biochemical processes at higher temperatures – open opportunities for study of thermophiles' role in cycle of matter in nature, especially in utilization of wastes of industry and agriculture [1].

* Institute of Microbiology, 7b A.Kadyri Street, 700128 Tashkent, Uzbekistan, E-mail: imbasru@uzsci.net; fax: 998-71-1442582

Earlier we have reported that mycelial fungi *Penicillium atrovenetum* and *Trichoderma harzianum* and yeast *Saccharomyces cerevisia* at their cultivation on solid and liquid media produce high content of protein and full complex of amino acids [2, 3].

The aim of the current study was isolation from nature of thermophilic micromycetes-producers of cellulases and fodder protein on base of assimilation of agricultural wastes and study of their protein synthesizing ability, and comparative study of fractional and amino acid composition of biomass protein of the most active fungal strains as well.

MATERIALS AND METHODS

The objects of study were thermophilic micromycetes isolated from water samples taken from different thermal springs of Uzbekistan (Akhangaran, Chirchik and Zarafshan regions) that were provided by Institute VodGeo.

Isolation of thermophilic cellulose destructing fungi from abovementioned samples was conducted by direct inoculation and by methods of repeated serial-dilutions method on surface of filter paper moisturized with nutrient medium and put on glass cannula in 250 ml Erlenmeyer flasks, and on surface of filter paper put on "starved" agar in Petri dishes as well.

Following acid media of Chastukhin modified by Zakharov [4]: $(NH_4)_2SO_4$ -0.75 g; KH_2PO_4 – 0.5 g; $MgSO_4{\bullet}7H_2O$ – 0.25 g; $FeSO_4$ – traces; $MnSO_4$ – traces; distilled water – 200 ml; pH 4.5-5.0 and physiologically alkaline solution of Chastukhin (4): KNO_3 – 3.0 g; KH_2PO_4 – 1.0 g; $MgSO_4{\bullet}7H_2O$ – 0.5 g; NaCl – 0.5 g; $MgSO_4$ – traces; distilled water – 200 ml; pH – 7.5-8.0 were used as nutrient media.

After inoculation Petri dishes were incubated at 40, 45, 50 and 55°C. Cultivation on abovementioned media promoted to clear differentiation of cellulose destructing fungi composition. The ability to synthesize cellulases and protein was determined in isolated pure cultures.

Cultivation during 5 days at 40, 45, 50 and 55°C was used for initial selection of fungi – cellulases and protein producers. It was conducted in 250 ml Erlenmeyer flasks containing 100 ml of Mandels [5] liquid medium with 2% of shredded filter paper. Two ml of suspension of conidia (density $1x10^6$/ml) were used for inoculation and cultivation was conducted on shakers with 200 rpm.

Influence of carbon sources on biosynthesis of cellulases and protein by selected active strains-producers was studied by cultivation on Mandels [5] medium containing of 2% of shredded filter paper or 2% of ground wheat straw or 2% of ground corn-cob as a single source of carbon.

To study influence of nitrogen sources on biosynthesis of cellulases and protein the selected fungal strains were cultivated on modified Mandels [5] medium containing 2% of ground wheat bran as carbon source and as nitrogen sources $(NH_4)_2HPO_4$, $NaNO_3$, NH_4NO_3, $(NH_4)_2SO_4$, urea and peptone were used in quantity of 0.2% by nitrogen.

Starting from 3rd day of cultivation, activity of cotton fiber hydrolyzing enzyme [6], activity of Na-carboxymethylcellulose (Na-CMC) hydrolyzing enzyme [7] and quantitative content of protein [8] was determined daily in culture broth. Content of reducing sugars was determined according to earlier described method [6]. The initial velocity of CMC hydrolysis equal to 1 mM of resulting reduced sugars in 1 min was taken as a unit of activity of enzyme. Content of fungal biomass was determined by dry weight.

Morphological and cultural properties of single-spore cultures of thermophilic micromycetes were studied by cultivation on Czapek medium with different sources of carbon, nitrogen and some growth activators.

Thermotolerant fungus *Thermomyces lanuginosus*, highly active protein producer, was cultivated on Mandels [5] medium with 2% sucrose for 3 days at 40°C on shaker at 220 rpm to study its biochemical composition. The quantity of inoculum was 1% to the volume of nutrient medium. 3-5 days old fungal mycelium was separated from nutrient medium by centrifugation, double washed with distilled water and used for research activities.

Five grams of dry biomass were homogenized by liquid nitrogen and repetitive freezing and melting for protein fractioning. The protein fractions were consistently extracted from cells homogenized by earlier described method [9]. Obtained protein fraction were dried in vacuum desiccator and in drying box at 50-60°C and then weighted. Taking weight of all fractions as 100%, the percentage of each fraction was calculated.

The quantitative content of amino acids in protein of biomass of *Th.lanuginosus* was determined according to method [10] on amino acid analyzer AAA-881.

RESULTS

To isolate new strains of thermophilic micromycetes water samples from thermal springs of Uzbekistan were used. Results of study showed that there is variety of different thermophilic micromycetes inhabiting local thermal springs. As result of screening about 100 fungal strains were isolated; 20 of them were often isolated and actively assimilated cellulose at high temperature of cultivation. These strains were studied more detailed.

According to obtained data (Table 1) the majority of thermophilic fungi (70%) was found in springs of Chirchik region that are characterized by high water temperature (55-70°C) and pH 7.5-8.0 compared to springs of Akhangaran and Zarafshan with lower water temperature (30-50°C). The most often met thermophilic fungi were *Aspergillus* and *Penicillium*. Temperature optimum for development of majority of them was about 40-50°C. Study of morphological and cultural features and biochemical properties allowed to identify them as *Aspergillus niger, A.terreus, A.fumigatus, A.flavus, A.ustus, Penicillium notatum, P.chrysogenum, Thermomyces lanuginosus, Torula thermophila, Thermoascus aurantiacus, Humicola grisea* and *Mucor michei*.

It is determined that thermophilic micromycetes grow on media with organic additions like corn steep, malt, autolyzed yeasts (Table 2). From carbon sources they assimilated glucose, sucrose, maltose, starch. Many of them used cellulose in form of filter paper or plant wastes. From nitrogen sources isolated strains assimilated mineral salts in form of $NaNO_3$ and $(NH_4)_2SO_4$ or their composition with organic sources.

It is necessary to note that 40°C was optimal temperature for cultivation of isolated strains that is why they are closer to thermotolerant forms. Only *Aspergillus fumigatus* and *Thermoascus aurantiacus* strains were able to grow and develop at 45-50°C. Thermotolerant microorganisms widely spread in nature and are actively used as producers for different preparations. But, the main positive feature of thermotolerant microorganisms that allows to identify this group from thermophilic and mesophilic microorganisms is high activity in wide temperature range from 30 to 50°C when velocity of their reproduction and activity remains intact.

Table 1. New strains of thermophilic micromycetes isolated from thermal springs of Uzbekistan

Strain	Location of spring, temperature and pH level of water	Number of thermophilic micromycetes to total number
Aspergillus niger-1c	Chirchik region; 55-70°C, pH 7-8	10
Humicola grisea-2c		5
Aspergillus terreus-3c		5
Aspergillus terreus-4c		20
Aspergillus fumigatus-5c		5
Aspergillus fumigatus-6c		5
Mucor michei-7c		7
Mucor michei-8c		3
Aspergillus flavus-9a	Akhangaran region; 30-50°C; pH 6-7	1
Torula thermophila-10a		3
Aspegillus ustus-11a		5
Aspegillus ustus-12a		1
Penicillium chrysogenum-13a		2
Penicillium chrysogenum-14a		3
Thermomyces lanuginosus-15z	Zarafshan region; 30-40°C, pH 5-7	5
Thermomyces lanuginosus-16z		1
Aspergillus niger-17z		2
Aspergillus niger-18z		3
Thermoascus aurantiacus-19z		2
Penicillium notatum-20z		2

c – Chirchik region; a – Akhangaran region; z – Zarafshan region

Due to stability at wide temperature parameters the thermotolerant forms of microorganisms are competitive in biotechnological processes. Comparative study of cellulase activity and productivity by protein and biomass of isolated thermophilic microorganisms revealed that all of them at different degree possess mentioned parameters (Table 3).

The highest cellulase activity was expressed by *H.grisea*-2c, *A.niger*-17z, *A.terreus*-4c and *A.ustus*-11a strains. Representatives of *Penicillium* genus had 2-3 times lower cellulolytic activity and *Thermomyces lanuginosus* strains were almost 10 times less active compared to maximum obtained data.

Table 2. Physiological properties of micromycetes isolated from thermal springs

Strains	Growth on media									
	1% corn steep	0,5% malt	2% autolyzed yeast	0,3% $NaNO_3$	0,5% $(NH_4)_2SO_4$	2% glucose	2% sucrose	2% maltose	2% starch	2% cellulose
A.niger-1c	++	+++	+++	+	+	++	++	+++	+++	++
H.grisea-2c	+++	++	+++	+	+	++	++	+++	+++	++
A.terreus-3c	+++	+++	++	+	++	+	++	++	++	+++
A.terreus-4c	++	+++	++	+	++	++	+++	+++	++	++
A.fumigatus-5c	++	+++	+++	++	+	++	+++	++	+++	+++
A.fumigatus-6c	+++	++	+++	+	+	++	+++	++	++	+++
M.michei-7c	++	++	++	++	+	+	+	+++	+++	+
M.michei-8c	+++	+++	+++	+	++	+	+	+++	++	+
A.flavus-9c	+++	++	+++	++	++	++	+++	+++	+	++
T.thermophila-10a	++	++	+++	+	+	++	++	+++	++	+++
A.ustus-11a	+++	+++	+++	+	++	+	+	++	+++	++
A.ustus-12a	+++	++	++	++	+	+	++	+++	++	+
P.chrysogenum-13a	++	+++	++	++	++	+	+	+	+++	++
P.chrysogenum-14a	++	++	+++	+	+	++	+	++	++	+
Th.lanuginosus-15z	+++	+++	++	+	+	++	++	++	+++	+
Th.lanuginosus-16z	++	+++	++	+	+	++	++	+	++	++
A.niger-17z	++	++	+++	++	+	+	+	+++	++	+++
A.niger-18z	++	+++	+++	+	++	++	+	++	+++	+++
Th.aurantiacus-19z	+	++	++	+	+	+	+	+++	++	+
P.notatum-20z	+	++	+++	+	+	+	++	++	+++	+

+++ - rich growth; ++ - good growth; + - weak growth

Biomass yield did not correlate with cellulase activity and the highest quantity of dry biomass was usually recorded at representatives of *Penicillium* genus. Biomass quantity at enzymatic active forms of *Aspergillus* was lower, apparently, in account of slower growth and development of cultures. But, protein content in filtrate of culture broth of *Aspergillus* and *Th.lanuginosus* strains was on high level.

Table 4 represents data on fractional composition of protein of freeze-dried biomass of investigated *Th.lanuginosus* strains. It is clear that yield of biomass protein in considerable extend depends on composition of solvent and phase of growth of culture. The highest quantity of biomass protein in studied strains is contained at exponential growth phase in water soluble fraction – 30.2-37.1%.

With increase of culture age the decrease of protein quantity in water soluble fraction at maximum stationary phase takes place and increase of its content in alkaline soluble fraction from c 18.0-19.3% to 26.1-27.1% from sum of soluble proteins.

Table 3. Cellulase activity and quantitative content of protein and biomass in cultural broth of fungi

Strains	Cellulase activity, u/mg	CMC-ase activity, u/ml	Protein, mg/ml	Biomass, g/100 ml
A.niger-1c	1.56	3.2	3.5	1.3
H.grisea-2c	2.64	6.0	3.63	1.3
A.terreus-3c	1.32	3.1	4.62	1.0
A.terreus-4c	2.44	5.3	3.63	1.2
A.fumigatus-5c	1.85	4.0	3.7	0.9
A.fumigatus-6c	2.54	5.5	3.78	0.5
M.michei-7c	0.56	0.9	4.1	1.5
M.michei-8c	0.43	1.1	4.38	1.9
A.flavus-9c	0.85	1.3	3.8	1.4
T.thermophila-10a	0.98	1.3	3.22	1.2
A.ustus-11a	2.54	6.2	3.4	1.4
A.ustus-12a	2.37	6.0	3.43	1.3
P.chrysogenum-13a	0.67	0.7	3.1	1.1
P.chrysogenum-14a	0.59	0.9	3.68	1.7
Th.lanuginosus-15z	0.32	0.3	4.75	1.5
Th.lanuginosus-16z	0.28	0.5	4.26	1.2
A.niger-17z	2.86	6.1	4.0	1.4
A.niger-18z	2.54	5.9	4.0	1.7
Th.aurantiacus-19z	0.88	1.2	3.22	1.8
P.notatum-20z	0.96	1.0	4.0	1.0

Data given in table 4 show that strain *Th.lanuginosus*-15z is high protein-producing culture, since there is the highest protein quantity in biomass in all studied fractions. Study of amino acid composition of protein fractions of biomass of *Th.lanuginosus*-15z revealed that sum of amino acids in albumin fraction is 32.45%, globulin – 19.23%, prolamine – 10.93% and glutelin – 9.24% out of total quantity of protein of biomass (Table 5).

Table 4. Fractional composition of biomass protein of Th.lanuginosus strains (% of total sum of protein)

Protein fraction	Growth phase					
	Exponential		Growth deceleration		Maximum stationary	
	Th.lanuginosus-15z	Th.lanuginosus-16z	Th.lanuginosus-15z	Th.lanuginosus-16z	Th.lanuginosus-15z	Th.lanuginosus-16z
Water soluble	37.1	30.2	32.1	30.0	25.1	23.1
Salt soluble	32.0	25.4	20.4	24.4	24.4	26.4
Spirit soluble	22.4	24.4	24.2	23.4	23.3	24.3
Alkaline soluble	18.0	19.3	22.0	22.0	27.1	26.1

Table 5. Amino acid composition of protein fractions of Th.lanuginosus-15z

Amino acid	Protein fractions			
	Albumin	Globulin	Prolamine	Glutelin
Lysine	1.08	0.88	0.52	0.34
Histidine	0.86	0.82	0.44	0.28
Arginine	2.82	1.04	0.86	0.48
Aspartic acid	3.26	1.72	1.04	1.02
Threonine	1.73	0.98	0.56	0.34
Serine	1.38	1.02	0.52	0.55
Glutamine acid	6.74	3.04	2.16	1.85
Proline	1.04	0.72	0.28	0.12
Glycine	1.02	1.04	0.57	0.44
Alanine	2.02	1.05	0.74	0.52
Valine	2.32	1.34	0.72	1.02
Cystine+ methionine	2.34	1.04	0.56	0.22
Leucine	1.08	1.08	0.84	0.58
Isoluecine	1.74	1.18	0.54	0.42
Tyrosine+ phenylalanine	3.02	2.28	0.58	1.06
Total sum of amino acids	32.45	19.23	10.93	9.24
Out of them:				
Essential	13.31	8.78	4.32	3.98
Contiguous	19.14	10.45	6.61	5.26

The prevailing contiguous amino acids are arginine, alanine, aspartic and glutamine acids. The highest quantity of essential amino acids is in albumin fraction – 13.31%, and the lowest is in glutelin fraction – 3.98% out of total quantity of protein in biomass. From essential amino acids the prevailing are valine, cystine+methionine, tyrosine+phenylalanine.

Study of amino acid composition of protein of biomass of *Th.lanuginosus*-15z revealed that content of aspartic and glutamine acids in total protein of biomass of *Th.lanuginosus*-15z increases with age of culture till beginning of the maximum stationary phase of growth and is equal, respectively, to 3.87 and 10.2% (Table 6).

Moreover, there is observed a repeatability of change in percentage of the metabolically most active amino acids like glycine, alanine, serine and valine in biomass protein of *Th.lanuginosus*-15z. The highest formation of abovementioned amino acids in biomass protein of *Th.lanuginosus*-15z takes place in exponential phase of growth.

The quantity of sulfur amino acids – cystine and methionine – increases in biomass protein starting from phase of growth deceleration and reaches the highest content in maximum stationary phase. It is noted that in maximum stationary phase of growth the increase of quantity of alkaline amino acids – arginine, histidine and lysine – takes place.

Thus, strains of thermophilic microorganisms that possess valuable properties (high cellulase activity and protein biosynthesizing ability) were isolated from thermal springs of Uzbekistan. It is shown that *Th.lanuginosus*-15z is more active protein-producing culture which contains high content of protein and full complex of essential and contiguous amino acids.

Table 6. Amino acid composition of biomass protein of Th.lanuginosus-15z out of total sum of protein

Amino acid	Growth phase		
	Albumin	Globulin	Prolamine
Lysine	1.08	1.24	1.36
Histidine	1.09	1.25	1.56
Arginine	3.04	3.45	3.84
Aspartic acid	3.35	3.65	3.87
Threonine	1.88	2.08	2.34
Serine	1.38	1.44	1.75
Glutamine acid	8.74	9.94	10.20
Proline	1.34	1.52	1.76
Glycine	1.24	1.44	1.66
Alanine	2.24	2.42	2.56
Valine	2.48	3.66	3.74
Cystine+ methionine	2.44	2.76	2.86
Leucine	1.19	1.32	1.49
Isoluecine	1.19	1.32	1.49
Tyrosine+ phenylalanine	3.22	3.42	3.76
Total sum of amino acids	35.90	40.91	44.24
Out of them:			
Essential	13.43	15.80	17.04
Contiguous	22.42	25.11	27.20

REFERENCES

[1] T.I.Bilay: *Naukova dumka*, Kiev, 1985, 171 p. (in Russian).

[2] J.Tashpulatov, B.G.Baybaev, T.S.Shulman: *Chemistry of Natural Compounds*, 4, 312-314 (2003). (in Russian).

[3] J.Tashpulatov, B.G.Baybaev, T.S.Shulman: *Chemistry of Natural Compounds*, 4, 315-317 (2003). (in Russian).

[4] M.A.Litvinov: *Nauka,* Leningrad, 1969, 169 p. (in Russian).

[5] M.Mandels, F.W.Parrish, E.T.Reese: *Bacteriol. J.*, 83 (2), 400-408 (1962).

[6] L.G.Loginova, J.Tashpulatov: *Applied biochemistry and microbiology*, 14 (3), 361-365 (1978). (in Russian).

[7] J.Tashpulatov, N.A.Teslinova: *Microbiology,* 18 (4), 609-612 (1974). (in Russian).

[8] O.H.Lowry, N.J.Rosenbrough, A.L.Farr, R.J.Randall: *Biol. Chem. J.*, 193 (1), 265-275 (1951).

[9] J.Tashpulatov, B.G.Baybaev, T.S.Shulman: *Chemistry of Natural Compounds*, 2, 209-211 (1998). (in Russian).

[10] V.I.Bilay (Ed.): Methods of Experimental Mycology, *Naukova dumka*, Kiev, 1982, 225p. (in Russian).

In: Biotechnology, Agriculture and the Food Industry
Editor: G. E. Zaikov, pp. 55-61
ISBN 1-60021-040-6

Chapter 9

THE EXPRESSION OF BACTERIAL *CSPD* IN TOBACCO ENHANCES THE RESISTANCE OF THE TRANSGENIC PLANTS AGAINST FUNGAL AND VIRAL PLANT PATHOGENS

K. A. Kromina and V. G. Dzhavakhiya*
Russian Research Institute of Phytopathology

ABSTRACT

Cold shock protein D from *Bacillus thuringiensis* induces nonspecific resistance in several plant species to viral and fungal infections. Several transgenic tobacco lines expressed *CspD* were obtained. Resistance of T0 and T1 transgenic plants to plant pathogens was tested in bioassays. Some transgenic tobacco lines were shown to possess an enhanced resistance to tobacco mosaic virus and the fungus *Alternaria longipes*.

Key words: *Cold shock protein, general elicitors, fungal and viral diseases, nonspecific resistance, transgenic plants*

INTRODUCTION

One of the promising approaches in plant protection is exploitation of natural plant defense potential. One of the strategies in plant protection is development of transgenic plants carrying an elicitor gene. There are some successful examples of expressing race-specific elicitors in plants resulted in increased resistance against pathogens [1, 2]. Race-specific elicitors usually cause HR in the host plant, which leads to an enhanced resistance against pathogens due to the microbe localization in the dead plant cells. A bottleneck of such approaches is constitutive expression of a race-specific elicitor gene that adversely affects the

* Russian Research Institute of Phytopathology 143050, Russia, Moscow region, Golitsino, Institute Street, e-mail: vitaly@vniif.rosmail.com or krominaks@yahoo.com, fax: +7 096 334 09 02

transgenic plant itself. To provide the normal plant growth, the gene of such elicitor should be controlled by an inducible promoter (e.g. wounding-inducible, pathogen-inducible). However, there remains a probability of the promoter leakage under unexpected environmental conditions. Another way to develop plants with enhanced resistance against plant pathogens is to express a general elicitor that does not trigger HR in the plant.

It was shown in our previous study [3], that low molecular (7.2 kDa) cold shock protein D (CspD was temporary named Microbial Factor 2[MF2]) from bacterium *Bacillus thuringiensis* induced nonspecific resistance in several plant species to viral and fungal infections. After treatment with elicitor, an increased resistance was observed for the following host plant – pathogen pairs: tobacco - tobacco mosaic virus (TMV), tobacco - potato virus X (PVX), potato – *Phytophthora infestans*, wheat – *Puccinia graminis*, rice – *Magnaporthe grisea*. The elicitor did not trigger HR in tested plants. There was found no adverse direct influence of CspD on fungal pathogens when added to spores. The gene of the elicitor was cloned, sequenced and its sequence deposited to GenBank (GenBank Accession No. AY272058). Later, another bacterial cold shock protein (CSP) from *Micrococcus lysodeikticus* (*Staphilococcus aureus*) was shown to possess an elicitor activity for *Solanaceae* plants [4].

The possibility of the transgenic plants development, carrying bacterial *CspD* gene, and resistance of such plants to plant pathogens were studied in this work.

Materials and Methods

Plant Material

Tobacco plants (*N. tabacum* cv. Xanthi NN) were propagated aseptically *in vitro* on a half-strength Murashige and Skoog basal salt mixture [5] supplemented with 30 g/l sucrose and 3 g/l phytagel (Sigma), pH 5.8 (½ MS). Tobacco plants were cultivated at + 24 °C with 16h light / 8h dark photoperiod.

Bacterial Strains and Plasmids

The supervirulent strain of *Agrobacterium tumefaciens* AGL0 [6] was used for transformation of tobacco plants by the standard leaf disk transformation method. For expression of *CspD* in plants the binary vector pBilt7 was constructed from pBin19 [7] by cloning *Hind*III fragments from pSS/*CspD* [derivative from pRT101 [8] into *Hind*III site of pBin19. The pSS/*CspD* (kindly provide by Dr. E. Kloke, Institute of Chorticultural crops, BAZ, Germany) contained the promoter and terminator of the cauliflower mosaic virus 35S transcript between which *CspD* was cloned into *Eco*RI/*Sma*I sites [9]. Plasmid pBilt7 was introduced into *A. tumefaciens* AGL0 [10]. This plasmid has the *nptII* gene fused to the nopaline synthase promoter and terminator together with *CspD* fused to CaMV 35S RNA promoter and terminator.

Agrobacterium-Mediated Transformation

Transgenic tobacco plants were obtained by the standard leaf disc transformation method using 100 mg/l kanamycin selection [11]. Transgenic plants were propagated on ½ MS medium with sucrose and then transferred to the soil.

DNA and RNA Isolation, PCR, Reverse Transcription PCR (RT-PCR)

DNA was isolated from 15 - 40 mg of plant tissue [12]. Primers for gene identification *CspD* were: forward "+1 *CspD*" and reverse "-178 *CspD*" (table 3). Each 12.5 μl of reaction contained 0.5 U of *Taq* DNA-polymerase (Fermentas), 10 mM Tris-HCl, 50 mM KCl, 1.5 mM $MgCl_2$, 200 μM of each dNTP, 5 pmol of each primer and 100 ng DNA-template. PCR was performed in a thermal cycler ("Terzic", Russia) with settings of 4 min at 94°C followed by 33 cycles of 1 min at 94°C, 40 sec at 56°C and 40 sec at 72°C and a final extention at 72°C for 5 minutes. PCR products were size-fractionated in a 1.2% w/v agarose gel, electrophoresis was performed in 0.5x TBE buffer [9].

Total RNA from leaves was isolated according to the RNA Isolation Kit manual (Stratagene®) and then treated with DNase I (Fermentas). First strand cDNA synthesis was performed with 1 μg of total RNA according to the manufacturer protocol with oligo-dT primer (Fermentas). RT-PCR was performed with gene-specific primers. First strand cDNA was used as template. To confirm the absence of genomic DNA in cDNA, PCR with gene-specific primers and total RNA treated with DNase I and not subjected to the first strand cDNA synthesis was performed.

Real-Time PCR Based on 5'-Nuclease Assay (Taq-Man Technology)

Primers and probes were designed with "Oligo" software (Table 3). Probes were labeled with 6-carboxyfluorescein (FAM) and Black Hole G1 (BHG1) as the quencher moiety (Syntol). Oligonucleotides for *CspD* were: forward primer "+88 *CspD*", reverse primer "-178 *CspD*" and probe 5' "+145 *CspD*". Oligonucleotides for the internal standard actin 9: forward primer "+427 *Nta-act9*", reverse primer "-484 *Nta-act9*" and probe "+453 *Nta-act9*". Each 25 μl of reaction contained 1 U of *Taq* DNA Polymerase, 10 mM Tris-HCl, 50 mM KCl, 4.0 mM $MgCl_2$, 200 μM of each dNTP, 12 pmol of each primer, 6 pmol of labeled probe and cDNA template synthesized from 2.5 ng of total RNA. PCR was performed in a SmartCycler II (Cepheid) with settings of 4 min at 94°C followed by 35 cycles of 15 sec at 94°C and 30 sec at 60°C. Fluorescence was measured at 60 °C. Additionally, PCR products were size-fractionated in a 2.0 % w/v agarose gel.

Bioassays with TMV

Transgenic and wild type tobacco plants [*N. tabacum* cv. Xanthi (NN)] were grown to the sixth leaf stage (for about 3 weeks) in pots with soil in a climatic chamber at RH (relative humidity) 60%, 24 °C the light and dark periods (12 h each). Detached tobacco leaves were rubbed with carborundum and then inoculated with TMV water suspension. The numbers of infective lesions were counted three days after the inoculation. Each experiment was performed 3 times with 6 plants per each line.

Bioassays with the Fungus *Alternaria Longipes*

For sporulation, *A. longipes* (kindly provided by Dr. R. Krämer, Institute of Chorticultural crops, BAZ, Germany) was grown on PCA medium [13] for 14 days at 20°C and under continuous UV light. Detached tobacco leaves (from plants of the sixth leaf stage) were evenly sprayed with a spore suspension of *A. longipes* (1×10^4 spores/ml) in 50 mM glucose with Tween 20. Inoculated leaves were maintained in Petri dishes contained wet filter paper and sealed with Parafilm tape at 20°C with 16 h light/8 h dark cycle. Lesions formed on leaves were scored in 7 days. Each experiment was performed 3 times with 6 plants per each line.

Results and Discussion

For *CspD* expression in plants, the binary vector pBilt7 was constructed from pBin19 plasmid [7]. Plasmid pBilt7 contained *nptII* gene fused to the nopaline synthase promoter and terminator together with *CspD* fused to CaMV 35S RNA promoter and terminator.

Tobacco leaf discs were transformed according to a standard method of *Agrobacterium*-mediated transformation. Kanamycin-resistant regenerants were screened by polymerase chain reaction (PCR) with *CspD* specific primers. mRNA from ten transgenic plants was subjected to reverse transcription (RT) - PCR to confirm *CspD* expression (Table 1). Several transgenic lines were lost due to the contamination during *in vitro* cultivation. There were no visible abnormalities in T0 plants' morphology and fertility in comparison with the wild type plants.

To evaluate the level of *CspD* expression, real-time PCR was performed. A house-keeping gene of actin 9 was used as an internal standard for cDNA quantification [14]. Transgenic lines X6, X10.6 (the line X10.6 was grown from a seed of transgenic line X10 after self-pollination because transgenic line X10 was lost) and X56 have shown similar level of *CspD* expression whereas its level in line X57 was definitely higher (Table 2).

Table 1. *Agrobacterium*-mediated transformation of tobacco

Number of explants	Number of rooted regenerants	Number of the transgenic plants	Number of regenerants with *CspD* expression at mRNA level	Transformation efficacy¶, %
173	57	10	7	5.8

¶ Transformation efficacy was determined as ratio of the transgenic plants number to the number of explants

Table 2. The level of *CspD* mRNA expression in the transgenic plants evaluated with real-time PCR

Transgenic line	X6	X10.6	X56	X57	Non-transgenic plant
Ct value for *CspD*	25.2	25.7	26.0	22.6	0
Ct value for actin 9	28.6	28.4	28.6	27.9	28.3

Table 3. Oligonucleotides used in PCR and real-time PCR

Gene name (GenBank Accession Number)	Oligonucleotide name (the number is position counting from the start codon)	Sequence
Cold shock protein D (AY272058)	+1 *CspD*	5' ATG CAA ACA GGT AAA GTT AAA TGG 3
	-178 *CspD*	5' TTA GTT TTT TGT AAC GTT AGC AGC 3'
	+88 *CspD*	5' TTC TCA GCT ATC CAA GGT GAC GG 3'
	probe +145 *CspD*	5' FAM - TTC GAA ATC GTT GAA GGT AAC CGT G – BHQ1 3'
Actin 9 (X69885)	+427 *Nta-act9*	5' CTT TTC CAA CCA TCA ATG ATT 3'
	-484 *Nta-act9*	5' CCA CAT CAC ACT TCA TGA TAG AGT T 3'
	probe +453 *Nta-act9*	5' FAM - GGA AGC TGC CGG AAT CCA CGA GAC T - BHQ1 3'

Transgenic lines X6, X10.6, X56, X57 and the non-transgenic tobacco were analyzed in bioassays with TMV and the fungus *A. longipes*. To obtain sufficient plant material the transgenic lines were propagated *in vitro*. Three-leaf tobacco were planted in the soil and grown for about 3 weeks before using in the bioassays.

Transgenic lines X10.6, X6, X56 were more resistant both to viral (Fig. 1) and fungal (Fig. 2) infections, whereas line X57 retained susceptibility to the plant pathogens at the level of the wild type plants. These results are consistent with our previous studies that have shown CspD to elicit defence responses in plants against both viral and fungal pathogens. The level of *CspD* mRNA expression for lines X6, X10.6, X56 was similar, which resulted in a similar pattern of resistance to the pathogens for these three lines. *CspD* gene was overexpressed in line X57. This line did not show any significant positive effect on the level of resistance as compared with the wild type plants. Probably, only a certain level of the elicitor in a plant results in an enhanced resistance against the plant pathogens.

Expression of the bacterial *CspD* gene could be used to address some plant pathology challenges.

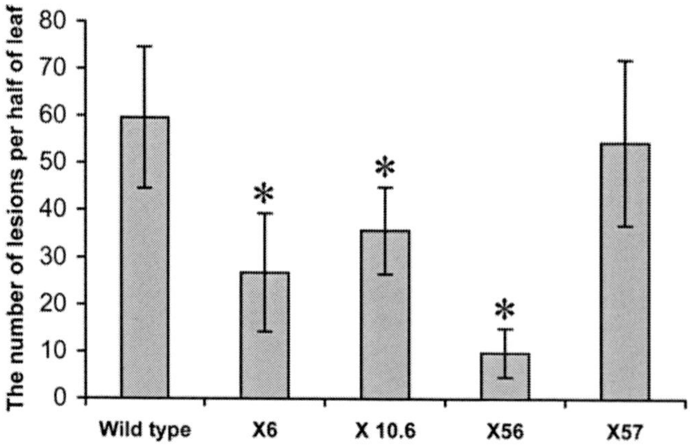

Figure 1. Estimation of resistance level in tobacco transgenic lines against TMV. The data shown are from typical experiment repeated three times with similar results. The data were analyzed with ANOVA statistics (P > 0.90), * shows the variant significantly different from a wild type plants. Bars represent averages ± SE of lesion number formed on leaves after inoculation with TMV for five leaves of individual plants.

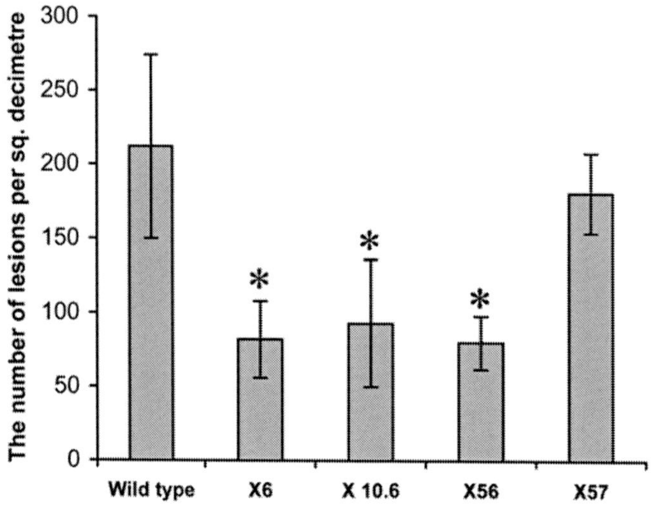

Figure 2. Estimation of resistance level in tobacco transgenic lines against fungus *A. longipes* The data shown are from typical experiment repeated three times with similar results. The data were analyzed with ANOVA statistics (P > 0.90), * shows the variant significantly different from a wild type plants. Bars represent averages ± SE of lesion number formed on leaves after inoculation with *A. longipes* for five leaves of individual plants.

REFERENCES

[1] Keller, H., Pamboukdjian, N., Ponchet, M., Poupet, A., Delon, R., Verrier, J.-L., Roby, D., Ricci, P. (1999) Pathogen-induced elicitin production in transgenic tobacco generates a hypersensitive response and nonspecific disease resistance. *The Plant Cell*, Vol. 11, 223–235.

[2] Honée, G., Melchers, L. S., Vleeshouwers, V. G. A. A., van Roekel, J. S.C., de Wit, P. J. G. M. (1995) Production of the AVR9 elicitor from the fungal pathogen *Cladosporium fulvum* in transgenic tobacco and tomato plants. *Plant Mol. Biol.* 29, 909–920.

[3] Djavakhia, V. G., Nikolaev, O. N., Voinova, T. M., Battchikova, N. A., Korpela, T., and Khomutov, R. M. (2000) DNA sequence of gene and amino acid sequence of protein from *Bacillus thuringiensis*, which induces nonspecific resistance of plants to viral and fungal diseases. *J. Russ. Phytopathol. Soc.* 1, 75-81.

[4] Felix, G., and Boller, T. (2003) The highly conserved RNA-binding motif RNP-1 of bacterial cold shock proteins is recognized as an elicitor signal in tobacco. *J. Biol. Chem.* 278, 6201-6208.

[5] Murashige, T., and Skoog, F. (1962) A revised medium for rapid growth and bioassays with tobacco tissue culture. *Physiol. Plant.* 15, 473-497.

[6] Lazo, G.R., Stein, P. A, Ludwig, R. A. (1991) A DNA transformation-competent *Arabidopsis* genomic library in *Agrobacterium*. *Biotechnology* 9, 963 – 967.

[7] Bevan, M. (1984) Binary *Agrobacterium* vectors for plant transformation. *Nucl. Acids Res.* 12, 8711-8721.

[8] Topfer, R., Schell, J., Steinbiss, H. H. (1988) Versatile cloning vectors for transient gene expression and direct gene transfer in plant cells. *Nucl. Acids Res.*, 16 (17): 8725.

[9] Sambrook, J., Fritsch, E. F., and Maniatis, T. (1989) *Molecular Cloning: A Laboratory Manual,* 2nd Ed., Cold Spring Harbor Laboratory, Cold Spring Harbor, NY.

[10] An, G., Ebert, P. R., Mitra, A. and Ha, S. B. (1988) *Plant Molecular Biology Manual A3*, (eds Gelvin, S. B. and Shilperoort, R. A.), Kluwer Academic Publishers, Dordrecht, 1988.

[11] Horsch, R. B., Fry, J. E., Hoffmann, N. L., Eichholtz, D., Rogers, S. G., and Fraley, R. T. (1985) A simple and general method for transferring genes into plants. *Science* 227, 1229-1231.

[12] Dorokhov, D. B., and Klocke, E. (1997) A rapid and economic technique for RAPD analysis of plant genomes. *Russian Journal of Genetics*, 33, No. 4, 358-365.

[13] Dhingra, O. D., and Sinclair, J. B. (1986) *Basic Plant Pathology Methods*, CRC Press, Inc. Boca Raton, Florida.

[14] Volkov R.A., Panchuk I.I., Schoffl F. (2003) Heat-stress-dependency and developmental modulation of gene expression: the potential of house-keeping genes as internal standards in mRNA expression profiling using real-time RT-PCR. *J Exp Bot.* 54 (391), 2343-2349.

In: Biotechnology, Agriculture and the Food Industry
Editor: G. E. Zaikov, pp. 63-68
ISBN 1-60021-040-6

Chapter 10

BIOTECHNOLOGICAL PRINCIPLES FOR PRODUCTION AND APPLICATION OF MICROBIAL PESTICIDES

E. I. Kolomiets ***and A .G. Lobanok***
Institute of Microbiology, National Academy of Sciences, Belarus

ABSTRACT

Theoretical basis has been laid for elaboration and application of biopesticides to protect cultivars from pests and pathologies. Approaches have been shown to isolate and select microbial antagonists and insect pathogens – promising agents for biological control of pests and pathogens. Mechanisms of phytoprotective action in most active cultures have been deciphered, parameters controlling biosynthesis of antimicrobial metabolites and a type of biotic interactions between antagonists and phytopathogens have been established.

Key words: *antagonists, antagonistic activity, actinomycetes, sporulating bacteria, biomethod, biopesticide, biotechnology, entomopathogen, insecticidal activity*

Necessity of research aimed at designing bilological control agents for plant bree ding is motivated by huge harvest losses induced by pests and diseases, on the one hand, and by sharp deficit of pathogen-suppressing agents, on the other hand.

The critical situation should be settled by launching large-scale manufacture of biopesticides readily fitting into existing plant protection systems owing to ecological safety and selectivity of action. Research and development of microbial pesticides envisages solution of the following key problems:

- engineering microbial strains distinguished by high antagonistic or insecticidal activity and the respective toxicity and pathogenecity tests;

* Institute of Microbiology, National Academy of Sciences, Belarus 220141 Minsk, Kuprevich str., 2. E-mail:microbio@mbio.bas-net.by, fax: (017)264-47-66

- investigation of mechanisms responsible for antagonistic or insecticidal action of potential biocontrol agents and tentative identification of antimicrobial metabolites;
- optimization of nutrient media composition and key factors regulating culture of microbial agents to be introduced into agrocycle;
- designing form of bioproduct, testing stability of active principle in storage;
- evaluating biological efficiency of designed compounds in laboratory, vegetation and field experiments;
- toxicity and sanitary expertise;
- compiling documentation package and state registration of biopreparation.

It appears natural, therefore, that our research efforts have been focused on isolation and comprehensive characterization of microbial antagonists and entomopathogens intended for use as agents for biological control of pests and pathogens. When screening promising variants, we were guided by such criteria as antagonistic/insecticidal activity, competitive ability in microbial cenoses, technological flexibility, safety for human health. Soil, plant phylloplane and rhizosphere served as isolation sources for antagonists, whereas entomopathogenic species were mainly isolated from host insects.

Applied techniques (selective media, membrane chambers, temperature treatment) yield isolates with defined properties and specific taxonomic affiliation. Completed studies resulted in collection of microbial antagonists and entomopathogens comprising bacteria of genera *Bacillus, Pseudomonas, Streptomyces* and fungi of genera *Trichoderma, Paecilomyces, Beauveria*. Our microbial isolates match best foreign analogs in antagonistic activity. *Actinomycete Streptomyces griseoviridis* showing efficiency towards phytopathogenic microorganisms and pests appears especially attractive. The strain may readily fit into various technological processes, it utilizes a broad spectrum of substrates and displays elevated adaptive capacity in biocenoses. It was found that antagonism of *S.griseoviridis* to phytopathogenic fungi is mainly caused by polyene antifungal antibiotics referred to pentaene group (fig. 1). High antibacterial activity of the actinomycete is determined by synthesized pigment complex. Conditions for submerged culture of *S. griseoviridis* were optimized, dynamics of culture growth and antifungal activity was analyzed. Based on obtained findings and screened isolate *S. griseoviridis*, technology of manufacturing biopesticide Biactin incorporating submerged actinomycete culture, separation of native antibiotic solution and composting of mycelial waste with lignocellulose substrates and dung was developed (fig. 2). Possibility for rational utilization of spent mycelium as lignocompost component ensures waste-free technology and ecological safety of the process. The resultant lignocompost product shows growth-stimulating and plant-protective effect, hence its application as fertilizer raises grain yields by 15 %. Liquid biopesticide Biactin effectively controls phytopathogenic microorganisms and pests.

Sporulating bacteria *B.subtilis* were engaged to design technology of producing biopesticide Fruitin untended to shield berry and fruit varieties from canker (pathogens *Nectria galligena*, *Pseudomonas syringae*), scab (*Fusicladium dendriticum*), gray rot (*Botrytis cinerea*).

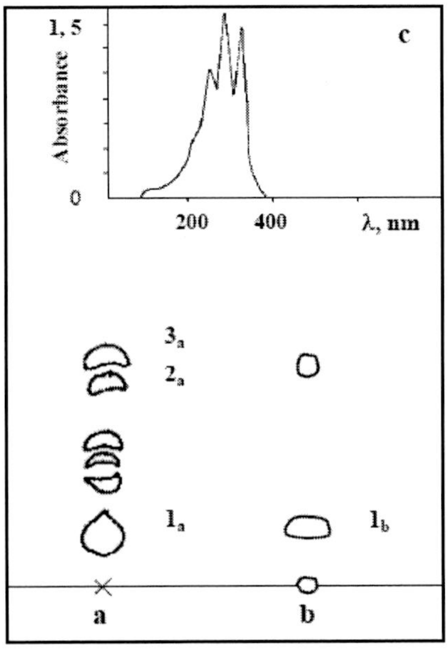

Fig. 1. TLC of ethylacetate cultural filtrate (a) and methanolic extract of *S. griseoviridis* mycelium (b), absorbtion spectra (c) of fraction with antifungal activity. 1_a, 1_b - fractions antagonistic towards *F. oxysporum*; 2_a, 3_a - fractions antagonistic towards *P. syringae*

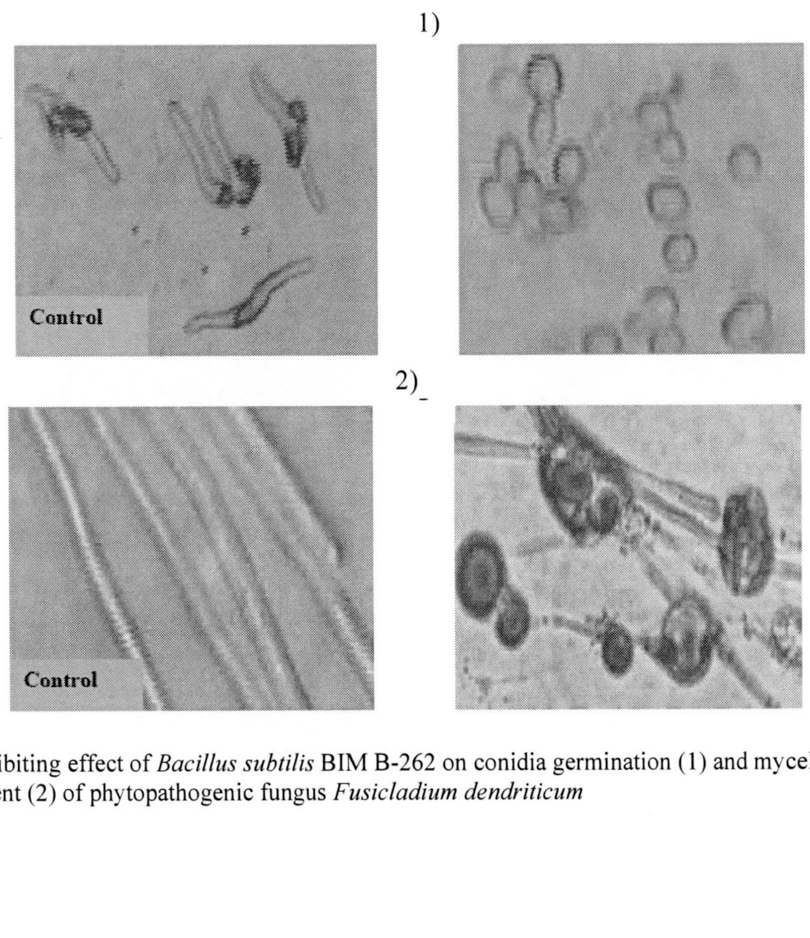

Fig. 2. Inhibiting effect of *Bacillus subtilis* BIM B-262 on conidia germination (1) and mycelium development (2) of phytopathogenic fungus *Fusicladium dendriticum*

The process envisages submerged bacterial culture on media with molasses–sugar refining waste. Economic efficiency of the scheme is guaranteed by relatively low duration (36-47 h) and waste recycling. The strain employed as active principle of Fruitin is characterized by increased antagonistic activity and growth rate, capacity to utilize a wide spectrum of sugar- and starch – containing substrates, which lifts its competitive potential and improves prospects for selection as biological control agent. Examination of biotic interactions between the tested antagonist and phytopathogenic fungi indicated that bacterial action led to mycelium deformation of scab pathogen fungus *F. dendriticum* and 88-95 % inhibition of conidial germination. Germinated spores either yield single deformed hyphas resulting in loss of reproductive function or completely cease functioning (fig. 3). The obtained results laid the basis for theoretically substantiated technology of Fruitin application. Since the biopecticide possesses marked capacity to inhibit germination of *F. dendriticum* spores, it should be introduced at periods when the pathogen passes conidial stage of development. Experiments in vivo supported this postulate. Integration of biopesticide into complex apple antiscab system slows down spread of the disease on fruit surface by 14,2-25,6 % as compared to the control, and yield of standard undamaged produce reaches 73,3-83,3 %. Biological efficiency of Fruitin for recovery of canker lesions in fruit trees ranges from 27,4 % to 51,6 %, exceeding effect of chemical agent Azophos taken as reference. Fruitin application reduces pesticide dosage in orchards and upgrades quality of produced fruits and berries. Biopesticide Fruitin successfully passed state registration procedure and was licensed for use in Belarus republic. Fruitin documentation package was developed and pilot-plant production of the product was launched at Novopolotsk bioconcentrate plant.

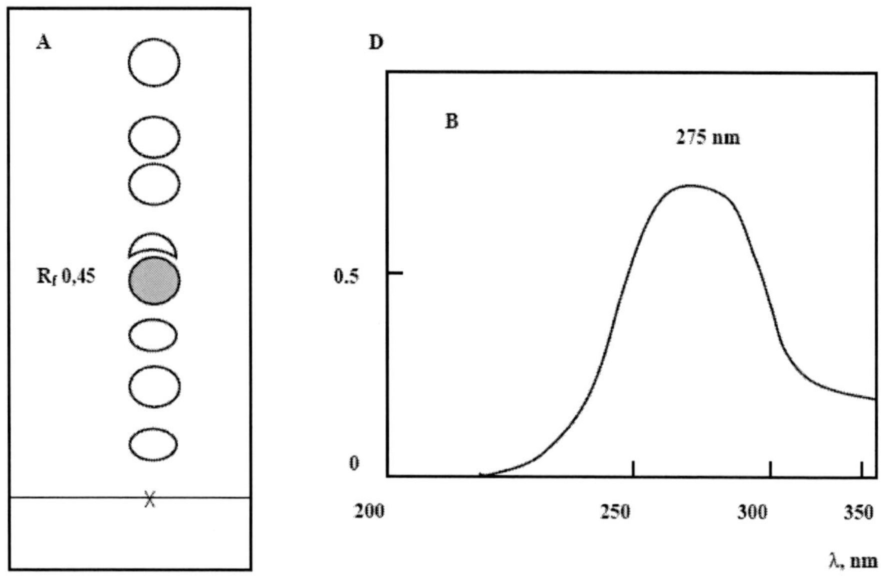

Fig. 3. Chromatogram (A) of methanolic extract of *B. pumilus* BIM B-263 supernatant on Silufol plates and absorbtion spectra (B) of fraction (R_f 0,45) with antibacterial activity

High efficiency in control of bacterial and fusarial infections severely damaging cultivars was demonstrated by isolated epiphytic strain of sporulating bacteria *Bacillus pumilus*. Submerged fermentation conditions were optimized for this culture so that the resultant antagonistic activity increased by 50-65 %. The tested strain is not inferior to recognized antagonists of genus Bacillus in ability to suppress development of phytopathogens *P. syringae* and *F. culmorum*. TLC results and chromato-mass spectrometry data indicate that high antibacterial activity of the culture is determined by a substance referred to aliphatic ketones-2,4-dimethyl-pentanone-3 (fig. 4). A series of similar compounds from this group distinguished by high antimicrobial activity is applied in medicine and agriculture to combat pathogenic species, for instance 1,3-dihydrooxy-2-propanone and its derivatives synthesized by bacteria of genera *Acetobacter, Aerobacter, Gluconobacter, Streptomyces.*

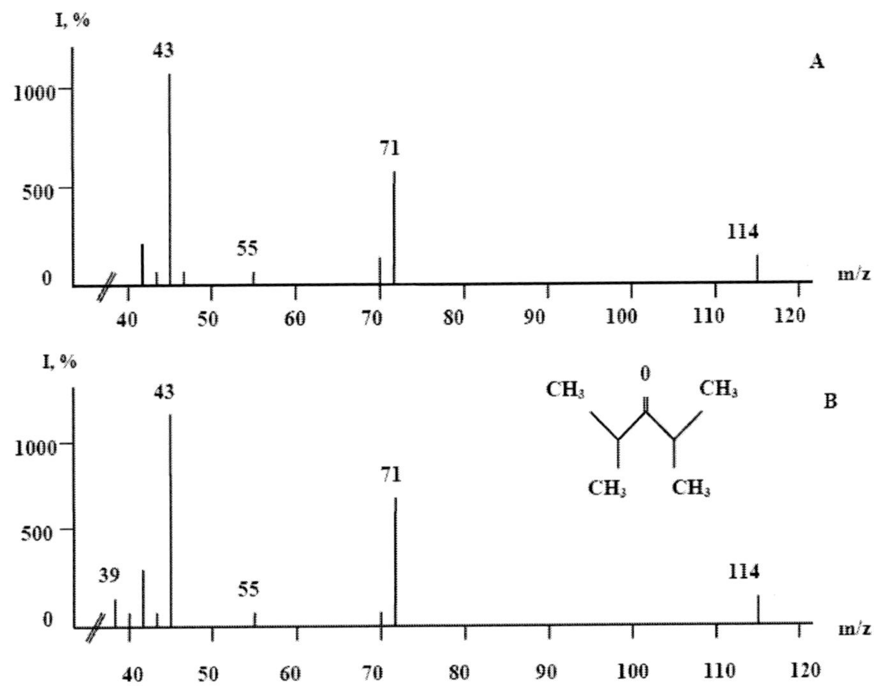

Fig. 4. Mass spectra of fraction R_f 0,45 with antibacterial activity isolated from methanolic extract of supernatant from *B. pumilus* BIM B-263 cultural luquid (A) and 2,4-dimethyl-pentanone-3 (B). Axis of abscissa – ratio of mass to charge, m/z; axis of ordinates – relative peak intensity (I), defined as the ratio of specific peak to maximum peak, %

Good commercial prospects are expected for our bioinsecticide Baciturin designed to control vegetable pathogens on greenhouse plots. Active principle of the product is β–exotoxin and spore-crystalline complex of bacteria *Bacillus thuringiensis*. We characterized cultural-morphological and physiological-biochemical properties of the culture (sporulation, generation of protein crystals, β–exotoxin, specificity to carbon and nitrogen sources, temperature optimum), defined factors regulating insecticidal activity. Obtained results contributed into technology of Baciturin production based on submerged culture of bacteria on readily available feedstock (barley meal, molasses). Upon registration trials bioinsecticide Baciturin was recommended for use against a broad range of greenhouse and field pathogens. Biological efficiency of the product in control of spider mite, carrot leaf flea, colorado beetle reaches 80-100 % matching best world analogs (Novodor, Bitoxybacillin) at 1.5-2.0 fold lower cost. Baciturin was formally registered at State Chemical Commission and pilot-plant manufacture of the product was started at Novopolotsk bioconcentrate plant.

A vital stage in microbial pesticide technology is design of preparation forms upgrading physical-chemical characterists and facilitating preservation of major functions, promoting activity and time of action, ensuring protection from UV-rays and oxidation. Diverse preparative forms of biopesticides and insecticides have been developed: dry, wetting and stabilized powders, emulsions, microcapsules, pellets. Recent years have seen a clear trend to replace powder by more compact damp forms. Such approach alleviates environmental pollution problem and allows to preserve natural biota associations.

Important role in raising efficiency of biopreparations is played by various supplements, like promoters of adhesion stickers surface active agents, stabilizers, photoprotectors, etc. Right choice of ingredient and commercial form of biopreparations allows to realize fully potential of microbial strains-sources of bioproducts, to raise their efficiency taking into account specific cultivars, environmental niche, biological peculiarities of phytopathogens and pests.

Completed investigations proved economic and ecological significance of manufacturing microbial pesticides in the form of flowing paste with addition of substances facilitating preservation and adhesive properties of the products.

Thus, our studies laid theoretical and practical basis for engineering ecologically safe biopreparations to control plant pests and pathogens.

The abtained experimental findings support the assumption on enhanced phytoprotective potential of actinomycetes and sporulating bacteria, disclose mechanisms and determine key factors controlling antagonistic and entomocidal activity, which makes a weighty contribution into modern concept of biological control of pests and pathogens as a cornerstone for large-scale fabrication of home-made plant protection agents to save hard currency and deficient resources.

References

[1] T.V. Romanovskaya, N.V.Sverchkova, E.I. Kolomiets, A.G.Lobanok: Vesti NAN Bel., *ser. Biol.*, 1, 41 (2002).

[2] E.I. Kolomiets: Vesti NAN Bel., *ser. biol.*, 4, 45 (1999).

[3] N.M.Golyshin: *Plant protection*, 8, 50 (1992).

[4] A.V.Poskryakov, G.V.Benkovskaya, M.G. Migranov: *Agrochemistry,* 8, 114 (1995).

[5] J. Segy: *Methods of soil microbiology*. Kolos Publ, Moscow, 1983. 253 p. (in Russian).

[6] E.I. Kolomiets,. N.A. Zdor, T.V. Romanovskaya et al.: *Bul. Pol. Acad. Sci., Biol. Sci.*, 47 (2-4), 89 (1999).

[7] L.N.Grigorcevich, E.I.Kolomiets, O.V.Molchan: *Bul. Pol. Acad. Sci., Biol. Sci.,* 50 (2), 95 (2002).

[8] T.V. Romanovskaya, E.I. Kolomiets, N.A. Zdor et al.: *Bul. Pol. Acad. Sci., Biol. Sci.*, 50 (2), 105 (2002).

[9] M.V.Shternshis: Raising efficiency of microbiological insect pest control. *Novosibirsk Agr*. Univ. Publ, Novosibirsk, 1995. 193 p.

In: Biotechnology, Agriculture and the Food Industry
Editor: G. E. Zaikov, pp. 69-74
ISBN 1-60021-040-6

Chapter 11

CALLUS FORMING AND REGENERATIVE ABILITIES OF SUGAR BEET PSEUDOMONAS-RESISTANT CELLULAR LINES

J. V. Kolomiets *

National Agriculture University, Ukraine

ABSTRACT

Metabolites of phytopathogen Pseudomonas wieringae which produces vascular-parenchymatous disease of sugar beet, in various concentrations were brought in nutrient medium Murasige and Scoog. In a selection response in culture in vitro was received to pathogen resistant cellular lines and plants-regenerates which were tested on resistance in environments in vivo.

Key words: *metabolites, Pseudomonas wieringae, sugar beet, callus, plants-regenerates, resistance.*

Cellular selection of plants finds more and more ways of usage for the accelerated reception of a valuable initial material, steady to phytopathogen [6]. Cellular lines are received with the help of simple selective system that allows selecting corresponding clones. After a period of cultivation and processing of mutagen, cells are sown out into nutrient medium which contains one or another pathogen. Only the cells that have got resistance to the pathogen as a result of mutations or reorganization of the genome survive [5]. The usage of this method is the most perspective for the crops, for which the methods of mass regeneration from callus, suspension crops and protoplasts have been developed. Those are tomatoes, lucerne, clover, and potato.

The agriculture of Ukraine is greatly affected by bacteriosises of sugar beet activated by *Pseudomonas wieringae*. The most effective method of harrowing against bacteriosis is the usage of resistant sorts and hybrids.

* National Agriculture University, 15, Heroev Oborony St., Kyiv 03041, Ukraine. e-mail: julyja@zeos.net

Cellular selection on phytopathogens resistance is carried out on the background of pathotoxin or cultural filtrate [9]. Cultural filtrate contains complex mixture of chemical compounds: polysacchrides, lipopolysaccharides, peptides, terpenoids, and proteins. Lipopolysaccharide, which has toxic qualities, can be found in the cultural filtrate as extracellular endotoxin. Free endotoxin has all the biological characteristics of its analog found in external membrane of the corresponding microorganism; the only difference is that lipopolysaccharide is more active [12].

The purpose of our research was to use the cultural filtrate of bacteriosis *Pseudomonas wieringae* agent as a selective factor for the selection of resistant cellular lines of sugar beet and to get plants-regenerates from resistant callus lines.

Methods and Materials

The object of research was callus tissues of 12 genotypes of sugat beet (Beta Vulgaris L.): sorts Belocercovskaja single-seed 45, Jaltuschkovskaja single-seed 64; diploid hybrids: Ukraininan MS 70, Verchnjatskiy MS 63, Ulado-Verchnjatskiy MS 37, Jaltuschkovskiy MS 72; triploid hybrids: Belocerkovskiy MS 57, Alexandria, Kaweros, Lena, Perla, Roberta.

Tootain callus lines of sugar beet we used nutrient medium of Murasige-Scoog (MS) [7]: MS1 (2,0 mg/l NAA + 0,4 mg/l BAP + 2,5 mg/l of an ascorbic acid), MS2 (0,1 mg/l NAA + 0,02 mg/l BAP + 25 mg/l of an ascorbic acid). Callus is received after 2-3 weeks of cultivation in diffused light on MS1 and MS2 medium with 24-25°C temperature. Suspension crop was cultivated during 14 days in darkness, in liquid nutrient medium (2-3 g callus tissues in 60 ml of medium) at constant mixing on a shaker (90-100 rev/min) and temperature 25-26°C.

Culture *Pseudomonas wieringae* 7922 was received from the museum of alive cultures of the Institue Microbiology and Virology by D.K. Zabolotnogo. it was cultivated within 5 days in a potato broth on a temperature-controlled appliance for cultivation of microorganisms at temperature 28°C. For the selection we used cultural filtrate (CF) which was received by filtration of cultural liquids through membrane filters "Synpor" with 0,4 micron pores.

For metabolite allocation more sparing method based on extraction with 0,85 % solution of NaCl [11] was used. The preparation were checked on the activity in reactions on precipitation [1] and double fusion in an agar on Ouchterlony [8]. The quantity of carbohydrated was defined with phenol-sulfuric method, the quantity of proteins – according to Lawry method [4], the quantity of neutral monosugars – with the method of gas-liquid chromatography [10].

To obtain resistance cell of sugar beet we sowed suspension crop during logarithmic growth phase into agar-agar MS medium with variable CF concentration. The nutrient medium MS without CF was used as control one. After incubation the colonies with more than 1mm in diameter were counted. The weight of callus tissues during cultivation was defined with Kucherenko method, and others [3].

Resistant sprouts were received in modified nutrient medium MS: MS3 (1 mg/l IBA + 0,05 mg/l BAP + 1 mg/l GA + CF); MS4 (2 mg/l IBA + 0,05 mg/l BAP + 2,5 mg/l GA + CF); MS5 (0,1 mg/l IAA + 0,5 mg/l NAA + 2 mg/l GA + 0,1 mg/l BAP + CF); they were cultivated at temperature +25°C, illumination 4klk and 16-hour photoperiod.

The sprouts of one size and with well-developed leaves were selected for rooting and were planted into nutrient mediums: MS6 (0,5 mg/l NAA + CF); MS7 (0,5 mg/l GA + 2 mg/l IBA + CF); MS7 (2 mg/l IBA + CF).

The resistance to bacterial diseases was defined with false infections [2].

Results and Discussion

Liposaccharides (LPS) were allocated from cultural filtrate as they are a component of an external membrane and — play a significant role in pathological process development. The received preparations were characterized by the presence of carbohydrates, proteins, neutral sugars and were serologically active with secondary O-serum (tab. 1)

Table 1. Chemistry and serological activity of preparations LPS *Pseudomonas wieringae* 7922

Activities		*Pseudomonas wieringae* 7922
Chemistry	Protein, %	11,5
	Carbohydrates, %	25,5
	Rhamnose, %	68,0
	Glucose, %	12,0
	Galactose, %	7,0
	KDO	+
Serological activity	Titers in reactions precipitation (1:)	500000
	Quantity of lines in reaction DDA	2

Table 2. Sublethal concentrations of metabolites *Pseudomonas wieringae* 7922

Sort, hybrid	*Pseudomonas wieringae* 7922	
	CF, ml/l	LPS, g
Belocercovskaja single-seed 45	40	2,0
Jaltuschkovskaja single-seed 64	40	2,0
Ukrainskiy MS 70	20	1,5
Verchnjackiy MS 63	20	1,5
Uladovo-Verchnjackiy MS 37	20	1,5
Jaltuschkovskiy MS 72	20	1,5
Belocerkovskiy MS 57	40	2,0
Alexandria	20	1,5
Kaweros	20	1,5
Lena	30-40	1,8-2,0
Perla	30-40	1,8-2,0
Roberta	30-40	1,8-2,0

Selection on the resistance to *Pseudomonas wieringae* 7922 supposed the selection of selective metabolite concentrations where the vital function of the cell with different resistance to the given factor was observed. The definition of metabolite sublethal dozes was based on comparative studying of the effect of different concentrations on the suspension crop growth. The sublethal concentration CF for callus tissues of sorts and hybrids are in table 2. As LPS exit is small (100-150 mg on 100 ml), the further work was carried out using CF of the researched culture.

During the cultivation at the presence of the selective factor most of cellular colonies perished. During the experience the tendency to decrease in callus biomass with increase in concentration of toxin in nutrient medium was observed. In the control medium the average weight of callus was from 0,53 to1,64 g (tab. 3).

Table 3. Average increase callus masses of different genotypes of sugar beet

Sort, hybrid	Average mass of callus, g
Belocercovskaja single-seed 45	1,56
Jaltuschkovskaja single-seed 64	0,99
Ukrainskiy MS 70	0,88
Verchnjackiy MS 63	0,63
Uladovo-Verchnjackiy MS 37	1,21
Jaltuschkovskiy MS 72	0,58
Belocerkovskiy MS 57	0,88
Alexandria	0,54
Kaweros	0,98
Lena	1,64
Perla	0,97
Roberta	1,39

Table 4. An output of plants-regenerates of sugar beet on selective medium with sublethal concentration of metabolites

Sort, hybrid	Resistant cellular lines, p	Morphogenetic resistant callus lines, p	Resistant plants-regenerators, p
Belocercovskaja single-seed 45	22	16	34
Jaltuschkovskaja single-seed 64	22	17	32
Ukrainskiy MS 70	7	5	10
Verchnjackiy MS 63	6	4	10
Uladovo-Verchnjackiy MS 37	7	5	9
Jaltuschkovskiy MS 72	8	5	10
Belocerkovskiy MS 57	16	11	20
Alexandria	15	10	18
Kaweros	15	10	18
Lena	20	15	30
Perla	19	16	28
Roberta	20	14	30

The essential difference in decrease of callus gain was observed at concventration of 10-40 ml/l CF *Pseudomonas wieringae* 7922. In general the growth of callus tissues of relatively more resistant sugar beet genotypes was oppressed less than of susceptible genotypes. So, at concentration 20 ml/l the weight gain of callus of the sorts Belocercovskaja single-seed 45, Jaltuschkovskaja single-seed 64 decreased for 40-42% accordingly to the control; of diploid hybrids Ukraininan MS 70, Verchnjatskiy MS 63, Ulado-Verchnjatskiy MS 37, Jaltuschkovskiy MS 72 — for 76-78 %; of triploid hybrids Belocerkovskiy MS 57, Alexandria, Kaweros, Lena, Perla, Roberta — 62-64%.

The selection of resistant cellular lines was made with the method of rigid selection according to the scheme: after two-three passages of callus in the condition of toxic medium resistant calluses were replaced into the control medium (without CF) and cultivated for 6 weeks to preserve their regenerative abilities. Then callus was again cultivated in the medium with sublethal dose of CF for three passages. As a result the resistant callus lines of the researched sugar beet genotypes were received.

Separate morphogenetic calluses of all genotypes were able to regenerate plants. The highest ability to regenerate in selective environment was observed in single-seed 45, Jaltuschkovskaja single-seed 64, triploid hybrids Lena, Perla, Roberta (tab. 4). The received plants-regenerates were characterized by slow growth and long rooting.

The results of risogenesis studying in selective environment showed that CF oppresses the development of root system comparing with the control (tab. 5). The ability of plants-regenerates to form roots was defined by genotype both in the control and selective medium.

To find out if the resistance to *Pseudomonas wieringae*, is saved the received lines were infected by *Pseudomonas wieringae* cells which had been grown on a potato agar during a day. The result showed that adding CF into nutrient mediums led to resistance increase for 1-4 points (according to 9-point scale) comparing with the control. The lines that were received in the medium without the selective factor were characterized by identical resistance with the initial form.

Table 5. Risogenesis resistant plants-regenerates to *Pseudomonas wieringae* 7922 in vitro

Sort, hybrid	Medium	Length of sprouts, mm	Length of rootage, mm	Quantity a root/sprout
Belocercovskaja single-seed 45	MS6	38	5	8
Jaltuschkovskaja single-seed 64	MS7	40	4	10
Ukrainskiy MS 70	MS8	39	4	8
Verchnjackiy MS 63	MS8	38,5	5	3
Uladovo-Verchnjackiy MS 37	MS7	28,3	6	9
Jaltuschkovskiy MS 72	MS6	19,4	3	5
Belocerkovskiy MS 57	MS7	18,7	6	2
Alexandria	MS6	18	3	1
Kaweros	MS8	18,5	5	1
Lena	MS7	39,2	8	8
Perla	MS7	38,8	9	6
Roberta	MS6	39,3	5	6

Thus, the selection with sublethal concentration of CF *Pseudomonas wieringae* allowed allocating resistant cellular lines. Adding CF into selective mediums led to the increase of resistance of plants-regenerates on 1-4 points.

REFERENCES

[1] K.I. Beltjukova, M.S. Matuschevskaja, M.D. Kulikovskaja, S.S. Sidorenko: *Methods of research of agents of bacterial diseases of plants*. Kiev, 1968. 316 p. (in Russian).

[2] N.G. Ivanova: *Bacterial test*. Moskow, 1987. 32 p. (in Russian).

[3] L.A. Kucherenko, R.P. Maddumage, J.L. Guzhov: *Agricultural biology*. (3), 84 (1991).

[4] O. Lowry, N. Rosenbrough, A. Farr, R. Randall: *J. Biol. Chem*., 133 (1), 265 (1951).

[5] I.M. Marunenko, A.A. Kuchko, T.N. Oleynik: *Notification recommendations*. Kiev, 1991. 30 p. (in Russian).

[6] O.J. Mesenceva: *Selection and seed-growing*., (4), 59 (1990).

[7] T. Murashige, F. SkoOG: *Physiol. Plantarum*., (15), 473 (1962).

[8] O. Ouchterlony: *Progr. in Allergy*., 5 (1), 1 (1958).

[9] V.A. Sidorov: Biotechnology of plants. *Cells selection*. Kiev, 1990. 280 c. (in Russian).

[10] I.J. Zaharova, L.V. Kosenko: *Methods of study microbe polysaccharides*. Kiev, 1982. 192 p. (in Russian).

[11] G.M. Zdorovenko et. al: *Microbiol. J*., 43 (5), 65 (1981).

[12] G.M. Zdorovenko: *Microbiol. J*., 50 (4), 98 (1988).

In: Biotechnology, Agriculture and the Food Industry
Editor: G. E. Zaikov, pp. 75-80
ISBN 1-60021-040-6

Chapter 12

STUDY OF SOME BIOCHEMICAL PECULIARITIES OF *PSEUDOMONAS* PRESERVED AT THE COLLECTION

M. R. Shakirzyanova**, *G. D. Zolotilina,
J.J. Tashpulatov and T. G. Gulyamova
Institute of microbiology, Tashkent, Uzbekistan

ABSTRACT

A number of bacterial strains were isolated from rhizosphere of technical, cereal and legume plants growing on saline soils of Tashkent, Syrdarya regions and from rhizosphere of wild plants growing on extremely saline soils of Republic of Karakalpakstan. Three strains from isolated cultures that stimulate seed germination and growth and development of cotton were selected. These strains expressed antagonism towards *Verticillium dahliae* and *Fusarium oxysporum* fungal pathogens. Selected strains actively grow on media with NaCl content up to 20% and pH 10. Studied strains could be recommended for the development of biopreparations for the needs of agriculture.

Keywords: *fluorescent bacteria Pseudomonas, rhizoshpere, pathogen biocontrol*

Increase of cotton and wheat diseases caused by fungal infection brings to considerable loss of agricultural products in Uzbekistan. Currently the great attention is paid to use of rhizospheric bacteria in biocontrol of agricultural plants from microbial pathogens [1, 2, 3].

Bacteria from Pseudomonas genus synthesize a number of secondary metabolites with considerable antimicrobial action and are widely used as components of biological means of plant protection world wide.

Howell and Stipanovic [2] reported active antagonist *Pseudomonas fluorescens* pf-5 that had high antagonistic activity against agents of cotton wilt, that is widely spread in Central Asia. Seed treatment with cell suspension of antagonist or its antibiotics led to increase of cotton resistance towards diseases up to 28-71%.

* Institute of microbiology, 7b A.Kadyri street, Tashkent, Uzbekistan, 700128. E-mail: imbasru@uzsci.net; fax: 998-71-1442582

Smirnov and Kiprianova [3] selected bacteria *Ps.aeruginosa* and *Ps.fluorescens* active by spectrum and intensity of fungicide action and supposed that studied strains synthesize active fungicide compounds.

Dafago and Haas [4] reported examples of effective use of *Ps.fluorescens*, *Ps.putida*, *Ps.cepacia* against *Fusarium sp.*, *Rh. solani*, *Erwinia sp.*, and other pathogens.

Mehrani et al [5] studied mechanism of fungicide action of *Ps. fluorescens in vitro* and its efficiency *in vivo* on cotton. They discovered fungicide action related to formation of physiologically active compounds. The strain was considered as potential biocontrol agent for seed treatment.

Pal at al. [6] studied possibility of use of Pseudomonas spp. strain EM 85 as biocontrol agent of *Rh.solani* that causes cotton root decay. Strain produced fungicide compounds and was successfully used for control of cotton disease.

The aim of the present research was study of Pseudomonas bacteria isolated from rhizosphere of plants growing on saline soils and possibility of their use as biocontrol agents of Verticillium and Fusarium cotton wilt.

Material and Methods

Bacteria from Pseudomonas genus preserved at the collection of the Institute of Microbiology and newly isolated strains from rhizosphere of technical, cereal and legume plants growing on saline soils of Tashkent and Syrdarya regions and from rhizosphere of wild plants growing on extremely saline soils of Karakalpakstan were used in experiments.

Soil sample collection, isolation of strains, study of their morphological and physiological features and identification was done with use of classical microbiological methods [3, 7, 8, 9, 10]. Method of direct cultivation and methods of accumulation on different media [3] were used for isolation of Pseudomonas bacteria from rhizosphere. NaCl in concentration of 10, 15, 20 and 23% was added to nutrient media for isolation of halophilic Pseudomonas strains.

Antagonistic activity of bacteria towards Verticillium and Fusarium was determined by methods of agar blocks and perpendicular strokes [8, 11].

Effect of NaCl and pH on growth of preserved at the collection and newly isolated bacteria was studied in liquid nutrient media at pH level 8, 9 and 10 and salt concentrations from 10 to 23%. Bacterial growth was recorded by presence of film, dregs, sediments.

Influence of secondary metabolites of fluorescent bacteria on seed germination and growth and development of cotton seedlings were studied in laboratory and vegetation experiments in Petri dishes. Seed germination was studied in dynamics during 7 days and expressed in per cents of number of germinated seeds to number of not germinated. Sterile and infested soil in 0.25 kg plastic cups was used for experiments.

Cotton seeds before sowing were sterilized, washed with sterile water and then treated with bacterial suspension ($1x10^7$ spores/ml). 1.5-2.0 mg of fungal pathogens were added to 1 kg of soil.

RESULTS

Long-term studies show that the most active antagonists towards phytopathogenic fungi *V.dahliae* and *F.oxysporum* are found among rhizospheric bacteria of cotton and cereals, especially among fluorescent *Pseudomonas* [2, 3, 12, 13, 14].

During recent years data are obtained on growth stimulating and antagonistic action of *Pseudomonas* on agents of other plant diseases. *Ps.putida* strain BPT1 decreased loss of legumes caused by another fungal pathogen – Botrys cinerea [15]. *Pseudomonas* strains GRP3A and PRS9 exerted also growth stimulating action on development of corn [16]. *Ps.fluorescens* and *Ps.reactans* B3 strains were the most active antagonists of fungal pathogen *Rhizoctonia solani* [17].

One hundred fifteen soil bacteria were isolated in pure culture from rhizosphere of cotton, wheat, alfalfa and salt tolerant plants – keyreuk (Salsola orientalis), wormwoods (Artemisia turanica, Artemisia diffusa), saltwort (Salsola arbusula), tamarisk (Tamarix orientalis). Fifty six out of them are small mobile gram-negative bacillus which do not form spores and produce fluorescent pigment. They are obligate aerobes, oxidase positive, do not need growth factors and practically do not grow at lower pH. According to aggregate of studied features the isolated bacteria were identified as *Pseudomonas*.

Table 1 represents data on antagonistic activity of newly isolated strains and 29 collection cultures of *Pseudomonas* towards *Verticillium dahliae* and *Fusarium oxysporum*. Thirty nine out of 85 studied strains (52%) suppressed development of *V.dahliae* and 13 strains (17.3%) suppressed growth of *F.oxysporum*.

Newly isolated and collection *Pseudomonas* strains expressed antagonistic activity towards phytopathogenic fungi at different degree (Table 2). From presented data, it is clear that fluorescent bacterial *Ps.aeruginosa*-133; *Ps.fluorescens*-13; *Pseudomonas* sp.-303; 307; 305 and 320 strains which cause growth suppression of *V.dahlia* and *F.oxysporum* have the potential for study of possibility of practical application.

Further the influence of fluorescent bacteria on growth and development of cotton infected with fungi-pathogens was studied. The results of experiment on Petri dishes revealed that 46% of tested bacteria exerted stimulating action on seeds germination and another 28% strains suppressed and hampered seed germination.

It is remarkable that strains of one species exert different influence on seed germination. Thus, from 13 tested *Ps.aeruginosa* strains 8 stimulated, another 3 inhibited and another 2 did not affect cotton seed germination. Same data were obtained on 11 *Ps.putida* strains (4, 2 and 5, accordingly) and 5 *Ps.fluorescens* strains (4 had a positive influence and one suppressed seed germination).

Table 1. Antagonistic activity of *Pseudomonas* strains

Cultures	Total	Growth suppression	
		F.oxysporum	V.dahliae
Collection strains	29	7	3
Isolated from rhizosphere of cotton, wheat and legumes	44	27	6
Isolated from rhizosphere of salt resistant wild plants	12	5	4

Table 2. Antagonistic activity of fluorescent bacteria towards phytopathogenic fungi (average from three experiments)

Zone of suppression, mm	Strains of fluorescent bacteria		Selected active cultures
	number	%	
Verticillium dahliae			
0	36	42	-
1-5	16	18.8	-
6-10	9	10.5	Ps.aeruginosa 40, Ps.fluorescens 123, Ps.putida 43; Pseudomonas sp 251, 301, 302, 304, 326, 331
11-15	7	8.2	Ps.aeruginosa 39, Ps.putida 42, Pseudomonas sp 301, 306, 524, 533
16-20	4	4.7	Ps.fluorescens 13, Pseudomonas sp 192, 203, 320
21-25	3	3.5	Ps. aeruginosa 133, Pseudomonas sp 303, 307
Fusarium oxysporum			
0	62	72.9	-
1-5	6	7.0	-
6-10	4	4.7	Ps.fluorescens 13, Pseudomonas sp 301, 307, 533
11-15	3	3.5	Ps.aeruginosa 133, Pseudomonas sp 305, 320
16-20	-	-	-
21-25	-	-	-

Table 3 represents data on seed germination in dynamics after their treatment by fluorescent bacteria of *Pseudomonas* genus. It is shown that seed treatment by fungi-pathogens considerable hampered cotton seed germination.

Analysis of cotton seedling grown on artificially infested soil revealed that treatment of seeds by cultural broth of fluorescent bacteria considerably reduces development of *Verticillium* and *Fusarium* wilts (Table 4). Application of bacteria-antagonists allowed to decrease decay of seedlings from 13.6 to 3.0% at *Verticillium* wilt and from 17.0 to 6.2% in case of *Fusarium*. At the same time the number of decayed seedlings in control was 59.8 and 87.8%, respectively.

Bacteria stimulating seed germination were tested in laboratory vegetative experiment. Double treatment of soil with cultural broth of bacteria positively influenced on velocity of cotton development and its resistance to wilt (Table 5). The length of stalk of 20 days old plants reached 15-17.2 cm and they had well developed root system.

It was determined that wilt affected cotyledonous leaves of seedlings treated by suspension of *V.dahliae* were 88,0 % and leaves of 20 days old plants – 57%. Obtained data showed that additional seed treatment by cultural broth of fluorescent bacteria considerably decreases wilt infestation. The best results were obtained at cotton seed treatment with cultural broth of *Pseudomonas* sp 307 and 320 strains.

Table 3. Influence of secondary metabolites of fluorescent bacteria on cotton seed germination in sterile soil (average from three experiments)

Variants	Seed germination time-course, days						
	1	2	3	4	5	6	7
Seeds soaked in sterile water (control)	-	9.4	16.3	32.2	64.1	74.2	83.0
Seeds soaked in cultural broth of Pseudomonas	-	12.5	25.7	51.4	75.2	87.3	92.8
V.dahliae infested seeds	-	-	5.9	14.8	41.6	68.0	80.1
F.oxysporum infested seeds	-	-	4.0	10.7	29.2	62.8	75.3

Table 4. Influence of fluorescent bacteria on infestation of cotton seedlings by phytopathogens on 7th day (average from three experiments)

Variant	Seedlings wilted in soil infested by V.dahliae, %	Seedlings wilted in soil infested by F.oxysporum, %
Seeds soaked in sterile water	59.8	87.8
Seeds soaked in cultural broth of:		
Ps.fluorescens 13	13.6	20.0
Ps.aeruginosa 133	3.0	17.0
Ps.putida 42	17.6	87.8
Pseudomonas sp. 192	17.0	87.8
Pseudomonas sp. 303	10.3	11.3
Pseudomonas sp. 305	59.8	11.3
Pseudomonas sp. 320	59.8	6.2
Pseudomonas sp. 301	59.8	19.3

100% - total number of germinated seeds

Table 5. Influence of cultural broth of fluorescent bacteria on growth velocity and resistance to cotton wilt (average from three experiments)

Variant	Wilt infestation, %		Height of stalk of 20 days old seedling, cm	Length of root of 20 days old seedling, cm
	Cotyledonous leaves	20 days old seedling's leaves		
Water soaked seeds	25.0	18.0	14.1	11.9
Seeds soaked in cultural broth of V.dahliae	88.0	57.0	12.0	10.0
Ps.fluorescens 133	25.0	12.0	13.4	12.0
Pseudomonas sp. 303	11.6	2.0	15.0	13.1
Pseudomonas sp. 307	4.7	1.0	17.2	16.0
Ps.putida 42	6.0	1.5	16.0	15.6

There were selected active cultures from strains preserved at the collection and newly isolated from saline soils that possess antagonistic activity towards *V.dahliae* and

F.oxysporum and stimulating cotton seed germination. The studied strains could be recommended for use as biological means of plant protection.

REFERENCES

[1] J.Cook: *Phytopath.*, 75, p.25-29 (1985).
[2] C.R.Howell, R.D.Stipanovic: *Phytopath.*, 70, p.712-715, (1980).
[3] V.V.Smirnov, E.A.Kiprianova: *Naukova dumka*, Kiev, 263 p, 1990. (in Russian).
[4] G.Dafago, O.Haas: Soil Biochem., 6, p.249-291 (1990).
[5] Mehrani, Disphand et al (2001).
[6] K.K.Pal, K.V.Tilak, A.K.Saxena, R.Dey, C.S.Singh: *Microbiol. Res.*, 155 (3), p.233-242 (2000).
[7] I.P.Babyeva, G.M.Zenonova: Soil biology. MSU publishing house, Moscow, 336 p., 1989. (in Russian).
[8] N.S.Egorov (ed.): *Practicum Metabolism of Bacteria.* MSU publishing house, Moscow, 307 p., 1986. (in Russian).
[9] V.G.Mineev: *Practicum on agrochemistry*. MSU publishing house, Moscow, 240 p., 1989. (in Russian).
[10] J.G.Holt et al. (Eds): *Bergey's Manual of Determinative Bacteriology*. Mir, Moscow, 1997. 800 p. (in Russian).
[11] N.S.Egorov (ed.): Practicum. MSU publishing house, Moscow, 245 p., 1995. (in Russian).
[12] A.D.Garagulya, L.V.Babich, E.A.Kiprianova, V.V.Smirnov: *Microbiol. J.*, 50 (6), p.77-81 (1988). (in Russian).
[13] G.M.Bianchini, R.D.Stipanovic, A.A.Bell: *Agric. Food Chem. J.,* 47 (10), p.4403-4406. (1999).
[14] G.Berg, N.Roskot, A.Steidle, L.Eberl, A.Zock, K.Smalla: *Appl. Environ. Mocrobiol.*, 68 (7), p.3328-3338 (2002).
[15] M.Ongena, F.Duby, F.Rossignol, M.L.Fauconnier, J.Dommes, P.Thonart: Mol. *Plant Microbe Interact.*, 17 (9), p.1009-1018 (2004).
[16] A.Sharma, B.N.Johri: *Microbiol. Res*, 158 (3), p243-248 (2003).
[17] F.Faltin, J.Lottmann, R.Grosch, G.Berg: Can. *Microbiol. J.*, 50 (10), p. 811-820 (2004).

In: Biotechnology, Agriculture and the Food Industry
Editor: G. E. Zaikov, pp. 81-84
ISBN 1-60021-040-6

Chapter 13

BIOREGULATORS CIRCON AND EPIN-EXTRA PROVIDE ENVIRONMENTALLY FRIENDLY PROTECTION FOR MEDICINAL PLANTS AGAINST INSECTS, DISEASES AND WEEDS

N. N. Malevannaya, G. P. Pushkina and L. M. Bushkovskaya
Non-commercial research-and-production partnership "Nest M", Moscow
All-Russian institute of medicinal and aromatic plants (VILAR), Moscow

In modern medicine herbs have not been forgotten, rather, interest in them is increasing. At the present time herbs are found to be widely applied not only as sources of effective drugs but also as biologically active supplements and for the enhancement of the human environment [1].

Until recently many medicinal plants were found as wild flora. If the goal of wild plant species in evolution is to survive, for herb cultivars it is to adapt to agrocenosis, thereby retaining good commercial characteristics. However, cultivation of wild plant species causes a change of ecological conditions, which reduces plant tolerance to pathogens and diseases.

Investigations undertaken by VILAR have shown that the profit-loss of medicinal plant resources due to pathogens, diseases and weeds is potentially 20-60 % [2]. That is why the protection of medicinal herb cultivars against pathogens during cultivation is an important aspect of technology. However, medicinal plants are a particular group of crops for which the use of chemical pesticides is strongly limited. Especially for crops which are used for galenic preparations. Moreover, factors such as environmental degradation and phytosanitary instability of herb biocenosis demands new approaches for the chemical protection of plants [3, 4].

Increasing interest in the application of environmentally friendly preparations in medicinal plant propagation led us to search for growth-regulating and immuno-modulating compounds based on natural resources to improve plant protection, stimulate plant growth and maturation, and increase their resistance to negative environmental impacts and pathogens. Why is it necessary to use these preparations during cultivation of medicinal herb crops? First, irregular and prolonged germination, and slow growth during the initial stages of ontogenesis are characteristics of herb crops, and secondly, it is well-known that such plants are usually very sensitive and intolerant to infections and diseases, and are unable to compete with weeds.

Over the last few years great success has been achieved in this field of agricultural research in testing preparations with protective and promotional activities, including Cirkon and Epin-extra.

Circon is of great interest due to its ability to enhance plant growth and protective action in the defense of plants against different diseases [5, 6]. The active ingredient of the preparation is a blend of hydroxycinnamic acids derived from the plant *Echinacea purpurea.* Due to the unique antioxidant activity of these phenolic compounds, Circon is able to influence diverse metabolic pathways. It is well-known that a group of phenolic compounds, including hydroxycinnamic acids (the active ingredient of Cirkon), are involved in a wide range of ecological, physiological and biochemical functions in plant cells [7]. Cells which are able to adapt to various stress factors have an observed increase in the content of phenolic compounds compared to cells with a low viability and adaptive capacity. For example, the presence of cadmium in the callus culture medium of long-fibred flax increases the content of phenolic compounds but the response of wild type cells is weaker (a characteristic increase of 20%) than the response of the cultural variety cells (an increase of 30 – 200%), which was caused by their ability to adapt to the impact of abiotic factors [8].

Circon enhances plant immunity, increases tolerance to unfavourable environmental factors, and provides an acute protective effect against diverse phytopatogens, exhibiting antifungal, antibacterial and insecticidal activities.

Epin-extra is a growth-stimulating preparation with an antistress effect. The active ingredient of the preparation is epiBS. EpiBS is a brassinosteroid which is a new class of phytohormones, first found and derived from rape pollen (*Brassica napus*) in 1979 [9]. The ability of brassinosteroids at exceptionally low concentrations (10^{-12}-10^{-7} M) to stimulate the germination of seeds, plant growth and development, to enhance the resistance of plants to stressful conditions, and to increase the crop productivity characterizes them as efficient, environmentally friendly growth regulators, which find application in plant propagation [10, 11].

Epin-extra activates the effects of gibberellins, cytokinins, auxins, abscisic acid and hydroxycinnamic acids. It protects against reduction of crop productivity by correcting the level of phytohormones and participating in shock protein synthesis in plants during stressful conditions.

The seeds of *Valeriana officinalis, Calendula officinalis, Echinaceae purpurea, Digitalis lanata* and *Hedysarum alpinum* treated with growth regulators Cirkon and Epin-extra exhibited an increase in germinating capacity of 8-10 %. The most effective germination (14% and 17%) was observed when Cirkon was used for herb seeds which exhibited a low initial germinating capacity. In this case shoots appeared 2-4 days earlier than did those of the control, and branching increased by 12-28%.

Subsequent treatment of vegetative plants with growth regulators promoted the growth process and formation of a stronger anabolic system. Compared to the control, the growth of aerial parts increased by 30 - 34% for *Leonurus cardiaca*, by 28 - 32% for *Dracocephalum moldavica,* by 20 - 26% for *Calendula officinalis,* by 26 –34% for *Echinaceae purpurea,* and by 28 - 40% for *Digitalis lanata*. The size of the assimilate surface increased 1.8-2.0 times.

Due to use of the bioregulators elongation of root systems were also observed, in addition to the growth of aerial parts. Thus, when *Valeriana officinalis* was treated with Cirkon its root mass increased by 40-45%. In the case of *Valeriana* this factor is very important because the roots themselves are sources for medicinal technology.

Besides an increase in growth of medicinal plant cultivars, bioregulators promote physiological changes in plants and accelerate the processing of phenophases. Cirkon facilitated early flowering in *Calendula officinalis, Echinaceae purpurea, Dracocephalum moldavica, Leonurus cardiaca* and *Hedysarum alpinum*. The flowering phase was observed 4-8 days earlier compared to the control.

The positive impact of the preparations mentioned above on the growth and development of plants caused an increase in productivity. Compared to the control herb crop capacity of treated cultivars was enhanced by 18-57%.

One of the characteristics of medicinal plants is that they contain biologically active compounds, and these substances define the medicinal value of the herb sources. Cirkon increased the glycoside content – lanatoside C – in *Digitalis lanata* leaves by 37.5%, Epin-extra – by 6.3%. *Echinaceae purpurea* treated with Epin-extra had an increased level of hydroxicinnamic acids of 29.6%. Cirkon enhanced the carotinoid content by 44% in inflorescences of *Calendula officinalis,* Epin-extra – by 14%. *Dracocephalum moldavica* crops treated with these growth regulating compounds afforded an essential oil yield increase of 45 -51%.

It was mentioned above that medicinal plants subjected to agrocenosis are less tolerant to pathogens. The loss of profit in crops of *Digitalis lanata* and *Echinaceae purpurea* because of root rot affection (*Fusarium*) can reach 12-60%. Septoria diseases can reduce the productivity of the leaves of *Digitalis lanata* by 30-40%, and active compounds content in raw material by 25-50%. During the last few years mixtures of traditional fungicides and SAR (systemic activated resistance) -inducers have been used to protect medicinal plants against diseases. It has been shown that preplant treatment of seeds of *Digitalis lanata, Echinaceae purpurea* and *Panax ginseng* with the growth regulator Cirkon facilitated a chemical disinfectant rate decrease (Kolfugo-super and Fundazol) by two times. In this case, not only was a decrease in the seed infection and germ affection by pathogens observed, but also an activation of growth processes. The efficiency and duration of the protective action of a mixture of natural SAR-inducers and chemical fungicides was higher than when fungicides were used alone. This integrated use of bioregulators and fungicides leads to the formation of plants with well-cultivated aerial and root systems and enables an increase in disease resistance of plants at later stages of morphogenesis.

A single treatment of the vegetative plants *Digitalis lanata* against Septoria diseases with the bioregulators Cirkon, Epin-extra and the fungicide Tilt facilitated a decrease in disease development in crops by 46-47% which was almost equal to double that of the treatment with the fungicide alone. The quantity of saved yield was 21.4 – 24.5 centner/hectare which was higher than the control by 55-57%. Moreover, in seasons with a low level of disease development, these protective-promotional preparations restrain disease development thereby increasing the natural resistance of *Digitalis lanata* to Septoria diseases and can make the use of fungicides unnecessary.

Sometimes nodulating weevils (*Sytona spp*) cause significant damage to shoots of *Hedysarum alpinum.* Due to their extreme population and harmful action 40% of shoots can be affected, which therefore require special protection. Shoot affection in alpine Hedysarum when treated with the insecticide Karate and with the growth regulators Cirkon and Epin-extra was only 3.8% and 4.1%, respectively. It should be noted that in these cases *Hedysarum alpinum* had a more active growth, and plants were higher than the control and samples treated with Karate alone by 40-46%.

Therefore, the results of investigations have shown that treatment with these growth regulators increases the resistance of *Hedysarum alpinum* to pests and in that way can reduce the usage of insecticide Karate by up to 34%.

It is possible to exclude the treatment of alpine *Hedysarum* shoots with insecticides if the population of pests is not more than the threshold quantity and the seeds and vegetative plants are treated with growth regulators.

Weeding is a significant part of the technology of cultivation of medicinal herbs. Field observations with series of herbs have shown that seeds and vegetative plants treated with growth regulators Cirkon and Epin-extra compete more effectively with weeds and can reduce herbicide use by 30-40%.

Therefore, the bioregulators Cirkon and Epin-extra enhance the phytosanitary condition of crops, increase crop capacity and quality, and are helping to create new environmental friendly technologies for the protection of medicinal plants against pests and diseases.

REFERENCES

[1] Bykov V.A., Zhouchenko A.A. *Medicinal plants propagation* // M. 2000. p. 148-155.

[2] Bushkovskaya L.M., Pushkina G.P., Bukina N.V. *The plant protection under the agro industrial complex reforms: economics, efficiency, and ecology* // S-Petersburg. 1995. p. 112-113.

[3] Tuterev S.A. *Biological basis and ways of use of the induced immunity to diseases and pathogens* // S-Petersburg. p. 5-21.

[4] Susidko P.I. *AGRO XXI*. 1997. No1, p. 4-5.

[5] Malevannya N.N. *Cirkon – a new stimulator of growth and development of plants* // 6th International Conference "Regulators of plant growth and development in biotechnology". M. 2001. p. 111-113.

[6] Kashirskaya N.A. *Cirkon and the resistance increase of fruit cultivars to fungi* // 6th International Conference "Regulators of plant growth and development in biotechnology". M. 2001. p. 244-245.

[7] Zaprometov M.N. Essential functions of phenolic compounds in plants // *Russian Journal of Plant Physiology*. 1993. V. 40. № 6. p. 921-931.

[8] Goncharuk E.A., Alyavina A.K. The studying of changes in the content of phenolic compounds in callus cultures of wild and cultural types of long-fibred flax under cadmium impact // *6th Symposium on Phenolic Compounds*. M. 2004. p. 25-26.

[9] Grove M.D., Spencer G.F., Rohwedder W. et. al. // Nature. 1979. V. 281. p. 216-217.

[10] Prusakova L.D., Chizhova S.I. Role of brassinosteroids in growth, resistance and productivity of plants // *Agrochemistry*. 1996. №11. p. 137-150.

[11] Ikekawa N. Brassinosteroid. A new plant growth substance // *Farmachia*. 1990. V. 26. №6. p. 548-553.

In: Biotechnology, Agriculture and the Food Industry
Editor: G. E. Zaikov, pp. 85-90
ISBN 1-60021-040-6

Chapter 14

IDENTIFICATION ANIMAL BY-PRODUCTS IN FEEDSTUFFS

Małgorzata Natonek-Wiśniewska, Ewa Słot and Agata Żyga
National Research Institute of Animal Production
Department of Immuno i Cytogenetyki Zwierząt. ul. Krakowska 1
32-083 Balice / Kraków, Poland

ABSTRACT

The Bovine Spongiform Encephalopathy caused great losses to cattle breeders because of deaths of affected animals and the need to cull those suspected to carry the disease. The lack of vaccines and tests enabling the elimination of diseased animals from the herd has made prophylaxis one of the main tools in fighting the disease. During numerous studies exploring the way in which this epidemic spreads, it has been stated that feed mixtures containing animal meals pose the greatest threat. Now feedstuffs can't contain animal by-products and hydrolysed protein, which molecular weight is over 10 kDa. In present publication method of identification of animal material and hydrolysed protein in feeds was described. Animal material was identified with mtDNA. Hydrolysed protein – based analyses were made with SDS-PAGE, and mtDNA was analysed with PCR.

Key words: *BSE, PCR, SDS-PAGE, hydrolysed protein, MBM*

INTRODUCTION

Methods for identification of animal components in compound feeds have been the subject of many studies in recent years. This is due to the risk of BSE in animals consuming compound feeds that contain animal components. Despite the official ban, these components are frequently added as a valuable source of the most readily available protein to animals. Protein is most often added in the form of meat-and-bone meals (MBM) or protein hydrolysates. Regulations on the supplementation of protein hydrolysates specify that they must not be used if their mass exceeds 10 kDa. This is why it became necessary to develop a method identifying the size of the hydrolysates tested. Regulations governing the addition of meat-and-bone meals set out that they must not be added or can be added under certain conditions.

SDS-PAGE, or vertical electrophoresis in polyacrylamide gel using sodium dodecyl sulfate (SDS), is a procedure used to identify the size of protein hydrolysates. SDS acts as an anionic detergent that binds the analysed proteins, thereby giving them a negative charge and giving the same charge to mass ratio to all proteins. Charging of proteins by SDS is high enough, so their actual charge is not important. This is why they divide only on account of their mass. The mass of a hydrolysate can be determined by comparing it with appropriate standards.

Meat-and-bone meals can be identified based on their DNA, which after isolation is amplified in PCR reaction and its products are separated during agarose gel electrophoresis. As in the previous method, the size of the products obtained can be determined through comparison with the size marker. This test analyses the species origin of the meal, because starter sequences characteristic of a particular species are used for PCR reaction.

Objective

The objective of the study was:

- To elaborate a method to distinguish protein hydrolysates with the mass above 10 kDa from those with the mass below 10 kDa.
- To identify meals sent for analysis for the presence of bovine and porcine components.

Material and Methods

The following samples were analysed:

1) two protein hydrolysates with the mass below 10 kDa
2) α-casein
3) BSA (Bovine Serum Albumin)
4) compound feed in which traces of bovine meat-and-bone meal were identified
5) compound feed in which traces of porcine meat-and-bone meal were identified.

The following procedures were used to identify protein hydrolysates:

1) Protein extraction technique, Methods used:
    (a) 100 g of sample was mixed with 100 ml PBS buffer, homogenized and centrifuged at 10 000 x g at 4°C for 15 min. The resultant supernatant was used for further analysis (acc. to Necidov et.al., 2002)
    (b) 0.5 g of compound feed containing the hydrolysate was dissolved in 1 ml H_2O, homogenized, left for 4 days, mixed in Vortex apparatus and centrifuged; the

resultant supernatant was diluted (1: 1000) in PBS buffer (acc. to Necidov et.al., 2002)

(c) 0.5 g of compound feed containing the hydrolysate was dissolved in 1 ml PBS buffer, homogenized, centrifuged, and the precipitate was used for further analysis (acc. to Ansfield, 1994)

(d) 0.5 g of compound feed containing the hydrolysate was frozen at -70°C, rinsed in 0.9% NaCl and homogenized for 20 s. The resultant supernatant was used for further analysis (acc. to Smidt et al, 1999)

(e) 0.5 g of compound feed containing the hydrolysate was dissolved in 0.9% NaCl and shaken for 1 min, followed by centrifugation. The resultant supernatant was used for further analysis (acc. to Smidt et al, 1999)

(f) compound feed containing the hydrolysate was homogenized in 10 volumes of 1% SDS buffer at 85-90°, and centrifuged. The resultant supernatant was used for further analysis (acc. to Smidt et al, 1999)

2) Separation of the protein, which included: selection of appropriate concentration of gel, its content of SDS, glycerol, addition of mercaptoethanol (1–2 μl) to the analysed sample, and elaboration of electrophoresis conditions such as current voltage or duration of electrophoresis. For the electrophoresis, 4 % concentrating gel and 6, 12.5 and 18% separating gel was used.

The identification procedure of bovine and porcine meat-and-bone meal covered the following procedures:

1) Isolation of DNA from compound feed based on guanidine thiocyanate (GuSCN) buffer, suspension of silicon dioxide (SiO_2) and casein kazeiny (Boom et.al, 1999)..
2) PCR reaction in which the analysed sequence was amplified with the participation of thermally stable DNA polymerase in the presence of starter sequences (Lahiff et al, 2001), nucleotides and magnesium ions.
3) Agarose gel electrophoresis in which the reaction products migrate in the gel as a result of applied voltage, their migration route depending on the size of the product, which is determined by comparison with size marker of known DNA fragment size. The results are visualized in UV light.

RESULTS

As a result of applied voltage, the protein sample migrates in the gel and, depending on the molecule size, its fractions can be observed in the gel as bands. The mass of proteins contained in the hydrolysate is determined by comparison with the protein marker of known size of fragments.

The gel image revealed bands for samples of hydrolysates whose protein was extracted following the method of Smidt (f). This shows that the procedure is suitable for testing protein hydrolysates. Extraction performed with other methods did not reveal bands in the gel.

During the tests it was concluded that the best conditions for protein separation were obtained for separating gel at a concentration of 12.5%, with 10% SDS buffer and 2 μl mercaptoethanol.

The clearest separation was obtained for 30 mA voltage and 40 min electrophoresis. For hydrolysate samples, the bands obtained were situated above 10 000 Da, indicating that proteins had migrated further, so their mass was lower than 10 000 Da. The results were also verified using two proteins (α-casein and BSA) with known fragment sizes. They were used at a concentration of 1.25 mg/ml. The results are shown in Fig. 1.

Meal in compound feeds is identified based on a DNA product of 212 bp for porcine meal and of 271 bp for bovine meal. As a result of amplification with bovine starters of DNA isolated from the first sample, a product of 271 bp attesting to the presence of the bovine component in the meal was obtained, while amplification with porcine starters of DNA originated from compound feed no. 2 gave a 212-bp band, which attests to the presence of the porcine component in this sample (Fig. 2).

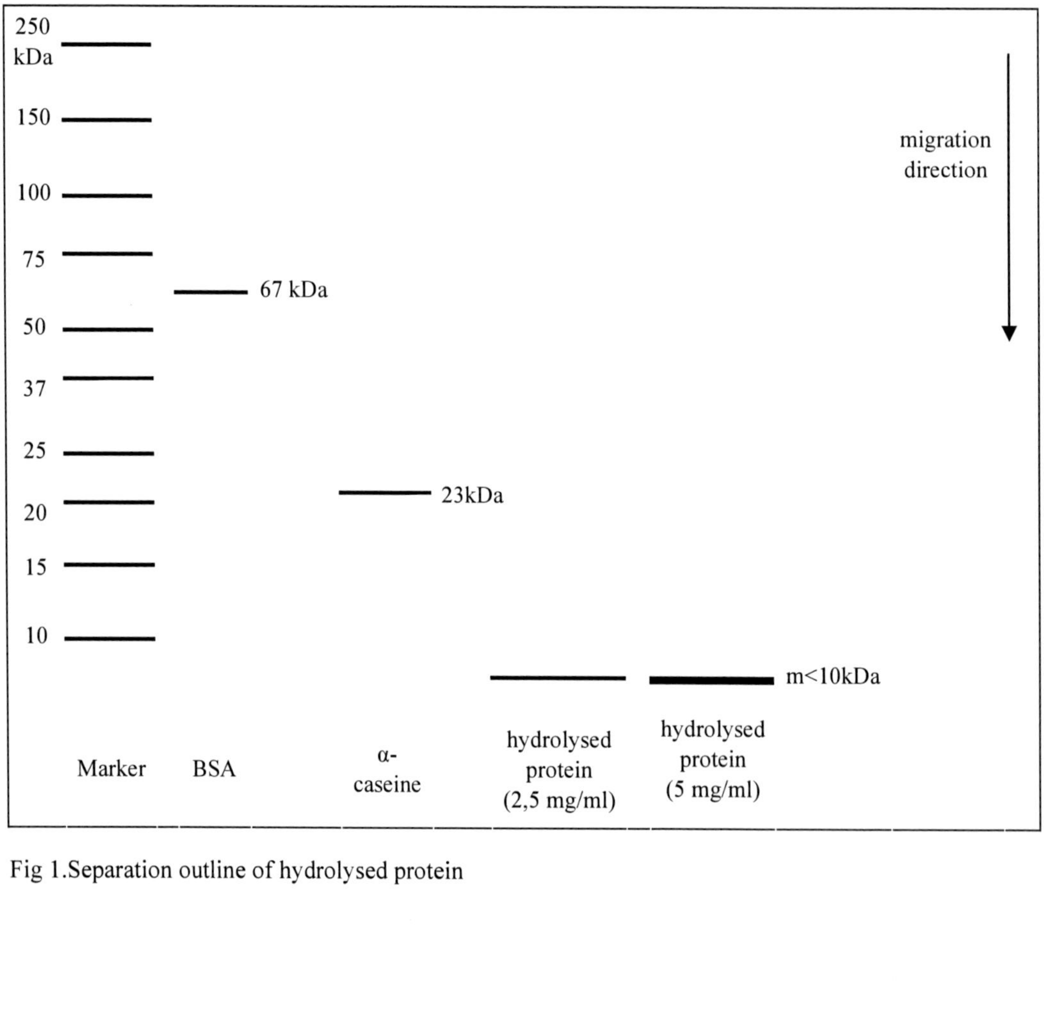

Fig 1.Separation outline of hydrolysed protein

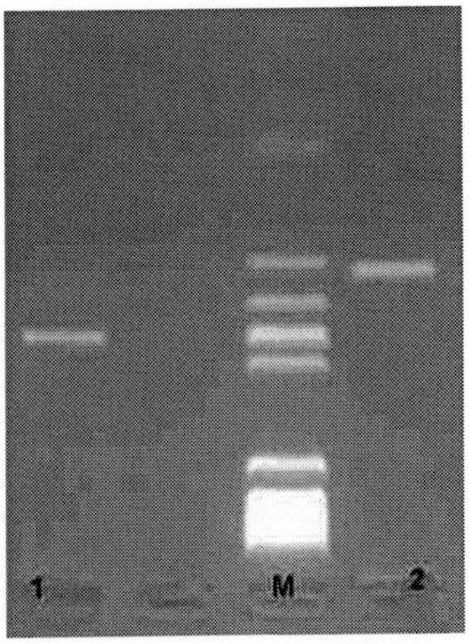

Fig. 2. Gel picture after electrophoresis sample including : 1) bovine MBM; 2) porcine MBM; M) PhiX 174 DNA/ Hae III ladder

DISCUSSION

The SDS-PAGE method allows for identification of molecules found in protein hydrolysates. This procedure determines the mass of polypeptide to an accuracy of 5-10%. The analysis of meat-and-bone meal provides quick information on provides quick information on the species origin of a component. The present study showed the results using the example of bovine and porcine components, but the described method can be also used to identify chicken and ovine meals. Both methods discussed in the present study allow for comprehensive identification of banned components that accounts for determination of species origin and the volume of protein supplements.

REFERENCES

[1] Ansfield M. 1994. Production of a sensitive immunoassay for detection of ruminant proteins in rendered animal material heated to >130°C. *Food Agric. Immunol.* 6: 419-433.

[2] Boom R., Cees S., Beld M., Weel J., Goudsmit J., van Dillen PW. 1999. Improved silica-guanidiniumthiocyanate DNA isolation procedure based on selective binding of bovine alpha-casein to silica particles. *J Clin Microbiol.* 37; 615-619.

[3] Lahiff S., Glennon M., O'Brien L., Lyng J., Smith T., Maher M., Shilton N. 2001 Species-specific PCR for the identification of ovine, porcine and chicken species in meat and bone meal (MBM). *Mol. Cell. Probes* 15(1): 27-35.

[4] Necidova L., Rencova E., Svoboda I. 2002. Counter immunoelectrophoresis: a simple method for the detection of species-specific muscle proteins in heat-processed products *Vet Med-Czech* 47(5).

[5] Schmidt G.R., Hossner K.L., Yemm R.S., Gould D.H., O'Callaghan J.P. 1999. An enzyme-linked immunosorbent assay for glial fibrillary acidic protein as an indicator of the presence of brain or spinal cord in meat. *J. Food Prot.* 62(4): 394-397.

In: Biotechnology, Agriculture and the Food Industry
Editor: G. E. Zaikov, pp. 91-93
ISBN 1-60021-040-6

Chapter 15

ANALYSIS OF *IGF2* GENE EXPRESSION IN PORCINE TISSUES

Maria Wrzeska**, *Agata Żyga and Barbara Rejduch

Department of Immuno- and Cytogenetics of Animals
National Research Institute of Animal Production
Balice, Poland

ABSTRACT

Expression of *IGF2* gene was analyzed in muscles, kidney and liver. As a marker of expression we used SWC9 microsatellite. Genotypes of chosen animals in SWC9 loci were established after isolation of genomic DNA from samples of blood according to the method described by Kawasaki with Coppiters`s modification and after PCR reaction. Expression of *IGF2* gene was analyzed by RT-PCR technique after isolation of RNA using Total RNA Isolation System *(Promega)*. Detection of RT-PCR product was performed by ABI PRISM 377 sequencer. Results of electrophoresis were analyzed in Genotyper 2.0 software. Expression profiles of SWC9 microsatellite in different tissues were compared with established genotypes.

Key words: IGF2, pigs, imprinting, RT-PCR, SWC9 microsatellite

INTRODUCTION

The *IGF2* gene is maternally imprinted in porcine fetuses (Jeon et al., 1999, Nezer et al., 1999, Kashiwazaki et al., 2003). It means that only the allele inherited from father is expressed in the offspring. The *IGF2* is one of the genes responsible for meatiness in pigs. Recently, a mutation in regulatory elements of the *IGF2* gene, responsible for phenotypic effect (muscle growth and BFT – backfat thickness), has been identified (Van Leare et al.,

* Department of Immuno- and Cytogenetics of Animals, National Research Institute of Animal Production, ul. Krakowska 1, 32-083 Balice, Poland. tel. 0048 12 2588313, fax 0048 12 2588150

2004) and it was shown that selection for *IGF2* may significantly improve meatiness in pigs. SWC9 is a microsatellite sequence localized in 3`UTR of *IGF2*. It was previously used as a marker of expression of the *IGF2* gene in a porcine fetuses (Nezer et al., 1999). The aim of our studies was to establish patterns of expression of the *IGF2* gene in porcine tissues of adult individuals.

MATERIALS AND METHODS

Samples of muscle, kidney and liver were collected from three heterozygotic pigs during the slaughter. RNA was isolated using SV Total Isolation System (Promega) from part of the samples (muscle, kidney and liver of one pig) immediately, the other part was frozen at -80°C. RT-PCR was performed using One Step RT- PCR System (Promega) with random hexamer primers and fluorescent labeled specific primers for SWC9 microsatellite. RT-PCR conditions were as follows: 48°C 45 min, 94°C 2 min, 94°C 13 min, (94°C 30 s, 58°C 30 s, 72°C 1 min)x35, 72°C 1h. The products were electrophoresed in a 4% poliacrylamid gel in ABI PRISM 377 sequencer. Results were analysed using genotyper w 2.0.

RESULTS AND DISCUSSION

After RNA isolation from fresh tissues and RT-PCR, we observed one allele in muscle and kidney, whereas in liver biallelic expression was detected (Fig.1). Studying frozen samples we did not observe any product in muscle tissue. In kidney we observed one allel and in liver we observed biallelic expression again. (Fig 2, 3). Our studies suggest that RNA should be isolated immediately from muscle, as it was difficult to obtain usable RNA from frozen tissue. Although, our results are preliminary and should be confirmed, they suggest that *IGF2* is imprinted in adult muscle and kidney, but not in liver. Nezer and Kashiwazaki observed monoallelic expression in liver, but they examined fetuses.

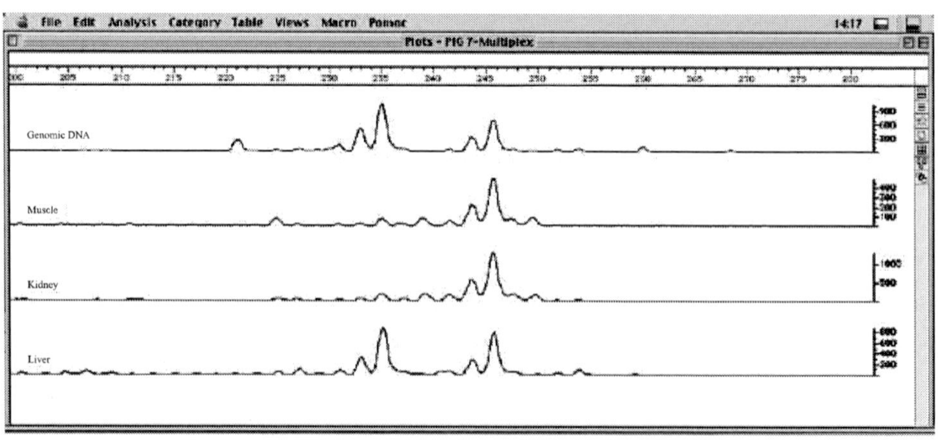

Fig. 1. Results of RT-PCR of SWC9 microsatellite (individual nr 1)

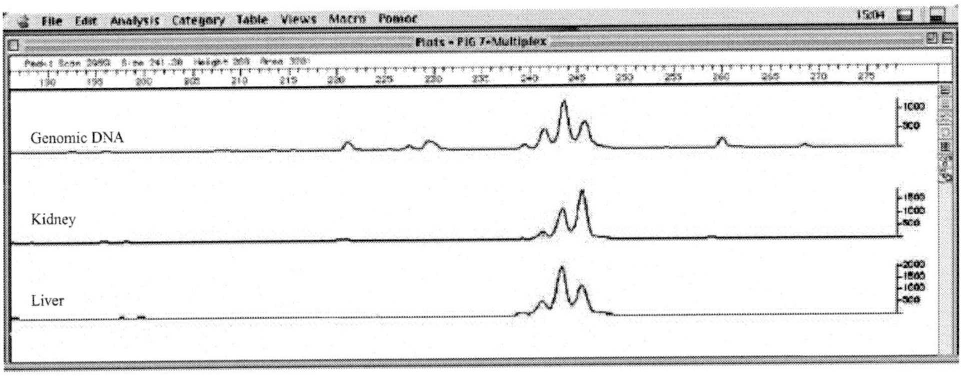

Fig. 2. Results of RT-PCR of SWC9 microsatellite (individual nr 2)

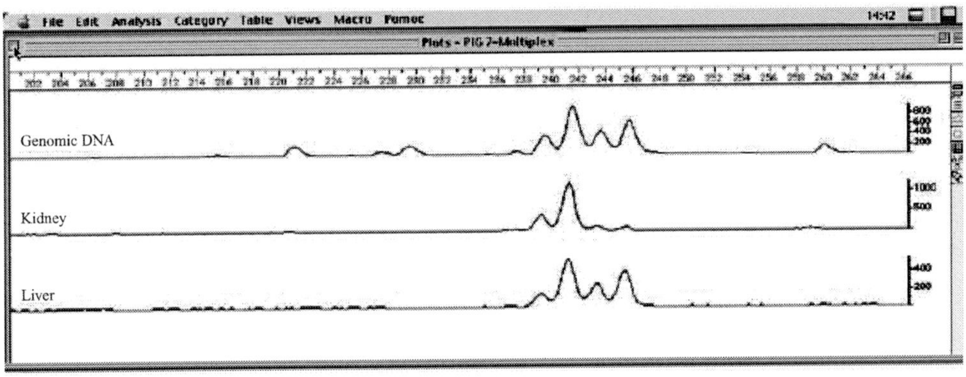

Fig. 3. Results of RT-PCR of SWC9 microsatellite (individual nr 3)

It appears that there is a loss of imprinting in adult liver. A Similar situation is observed in humans: *IGF2* is imprinted in most tissues but is expressed from both alleles in liver and choroids plexus (review: Morison and Reeve, 1998).

REFERENCES

[1] Jeon J. T, Carlborg Ö,. Törnsten A, Giuffra E, Amerger V., Chardon P., Andersson-Eklund L., Andersson K., Hansson I., Lundström K., Andersson L., (1999) *Nature Genetics* 21, 155-156.

[2] Nezer C., Moreau L., Brouwers B., Coppieters W., Detilleux J., Hanset R., Karim L., Kvasz A., Lercy P., Georges M., (1999) *Nature Genetics* 21, 157-158.

[3] Kashiwazaki N, Yamaguchi R., Hisamatzu S., Ichihara N., Asari M., Sonoki S., Kono T., Shino M., (2003). *Theriogenology* 59, 423.

[4] Morisonm., Reeve E., (1998). *Human Molecular Genetics* 7;10:1599-1609.

[5] Van Laere S., Nguyen M., Braunschweig M., Nezer C., Collette C., Moreau L., Archibald A. L., Haley C. S., Buys N., Tally M., Andersson G., Georges M., Andersson L., (2003) *Nature* 23;425(6960):832-6.

In: Biotechnology, Agriculture and the Food Industry
Editor: G. E. Zaikov, pp. 95-100
ISBN 1-60021-040-6

Chapter 16

DISTRIBUTION OF *BACILLUS THURINGIENSIS* IN THE AGRICULTURAL SOILS OF IRAN

***R. Marzban*[1] *and G. R. Salehi Jozani*[1,2]**
[1]Biological control Research Department,
Plant pests and Diseases Research Institute, Tehran Iran
[2]Vavilov Institute of General Genetics,
Russian Academy of Sciences, 119991 Moscow

ABSTRACT

Bacillus thuringiensis was isolated from 2'234 soil samples collected from different provinces of Iran. The results of this study clearly revealed that *B. thuringiensis* was distributed in the agricultural soil of over 80% of Provinces. A total of 128 crystal-forming isolates were successfully obtained from 28'445 bacterial colonies examined. In toxicity tests, 44% of all isolates was toxic to Lepidoptera, 21% was toxic to Coleoptera, 7% was toxic to Lepidoptera / Coleoptera and 28% was non-toxic isolates. Thirty-five all Lepidopteran-active isolates produced typical Bipyramidal inclusions, and the remainder, which belong to Coleoptera-active and Lepidopteran / Coleopteran-active isolates produced bipyramidal or cuboidal inclusion.

Keywords: *Bacillus thuringiensis, Isolation, Distribution, Agricultural soil, Iran*

INTRODUCTION

Bacillus thuringiensis (*Bacillaceae*) is a Gram-positive, rod-shaped, spore forming bacterium that produces insecticidal crystal proteins during sporulation. *B. thuringiensis* has been isolated from environments associated with insect population and plant material. Recent surveys indicate that *B. thuringiensis* is distributed in soil sparsely but frequently and its

[1] Biological control Research Department, Plant pests and Diseases Research Institute, P. O. Box 1454, Tehran 19395, Iran. Tel/Fax: +98 (021) 2420224, gsalehi2002@yahoo.com

distribution is widespread. Dulmage and Aizawa (1982) suggest that the normal habitat of *B. thuringiensis* is in the soil. Since then there have been many efforts for isolating new *B. thuringiensis* from soils of several Countries (De Lucca *et al.* 1981, Ohba and Aizawa 1986, Padua *et al.* 1982, Anwar Hossain *et al.* 1997, Bernhard *et al.* 1997, Kim *et al.* 1998). Martin and Travers (1989) reported isolation and distribution of *B. thuringiensis* from soil samples of five continents and their associated islands. They found that *B. thurngiensis* was the most abundant in the samples from Asia. According to Attathom *et al.* (1995) most of *B. thuringiensis* serotypes of Thailand produced bipyramidal shaped crystal. Marzban (2002) reported that native isolates were more effective than var. *kurstaki* (Dipel) on third instar larva of Indian meal moth (*Plodia interpunctella).* The aim of this study was to examine the distribution and insecticidal activity of *B. thuringiensis* isolates in agricultural soils of Iran that had never been treated with *B. thuringiensis* products.

Materials and Methods

Sample Collection

Samples of about 200 g soil, scraped from 10 cm depth with a sterile spatula, were collected randomly from agricultural lands in different provinces of Iran. The samples were placed in sterile plastic bags and stored at 4°C. The fields from which these samples were taken had not been previously sprayed with any commercial formulation of B. thuringiensis. Furthermore, since B. thuringiensis is rarely applied as an insecticide on crops in Iran, it is unlikely that commercial isolates of B. thuringiensis were present in our study.

Bacterial Isolation

Isolation of bacteria was done according to the modified method of Anwar Hossain et al. (1997). Two hundred g of sample were suspended in sterile physiological saline (1:5, w/v), mixed vigorously, and left to stand without any disturbance. A total of 2 ml of the clear supernatant was heated at 80°C for 5 min. The heat-treated supernatant was grown on nutrient agar plates at 31°C for up to 5 days. Colonies similar in morphology of bacteria were selected. Colonies were examined under a phase-contrast microscope for the presence of parasporal inclusion bodies and spores.

Insect Bioassays

B. thuringiensis isolates were grown on a nutrient agar plate at 31°C for 7 days until completion of sporulation. Tests for toxicity to Lepidopteran and Coleopteran insects were conducted at 25°C in 60-70% relative humidity with a 16 hr light-8 hr dark cycle. All tests were performed with a spore-crystal suspension of 3×10^8 CFU/ml. For Lepidopteran assay, 3 ml spore-crystal suspension were mixed into 10 g of artificial diet. One gram of the above-mentioned artificial diet was placed in a camera film tube and then a single 4-day-old

Helicoverpa armigera larva was put individually on the diet. Typically, 10 larvae were used per isolate. The mortality was recorded after 72 hr. For toxicity test against *Leptinotarsa decemlineata* (Say) young potato leaves were dipped into spore-crystal suspension of 3×10^8 CFU/ml, and placed in a Petri dish. After the surface of the leaves had dried, 10 second instars larvae were placed onto 3 leaves in each Petri dish. Larval mortality was recorded for up to 72 hr. Percent larval mortalities were corrected following Abbott's formula (1925).

RESULTS AND DISCUSSION

Table 1 shows distribution of *Bacillus thuringiensis* (*Bt*) in soil samples collected from agricultural fields of different provinces of Iran and the frequency of crystal-forming isolates in bacterial colonies examined. From the examination of a total of 2'234 samples which were collected from all provinces of Iran, it is clearly revealed that *B. thuringiensis* was distributed in the Agricultural soil of over 80% of Provinces. Furthermore, a total of 128 crystal-forming isolates were successfully obtained from 28'445 bacterial colonies examined. While the largest number of *B. thuringiensis* isolates was collected from the soil of cotton, seed oils, sugar beet filds and rice farms, the smallest number was isolated from the soil of vegetable farms. Martin and Travers (1989) studying the distribution of *B. thuringiensis* in 50 countries which represented all five continents of the world, found that the most abundant isolates of *B. thuringiensis* came from Asian countries.

Table 1. Distribution of *Bacillus thuringiensis* (*Bt*) in soil samples collected from Agricultural fields of different provinces of Iran and frequency of crystal-forming isolates in bacterial colonies examined (Isolates/Colonies)

Type of cultivation and location	Frequency of *Bt* in samples examined	Isolates/Colonies
Wheat field, Ilam	1 / 50	1 / 272
Rice field, Ilam	0 / 10	0 / 45
Vegetable field, Ilam	0 / 15	0 / 73
Wheat field, Kermanshah	4 / 61	4 / 127
Summer crops and Vegetable field, Kermanshah	5 / 65	5 / 1398
Sugar beet field, Kermanshah	1 / 50	1 / 161
Wheat field, Kordestan	6 / 56	7 / 1421
Summer crops and Vegetable field, Kordestan	0 / 40	0 / 70
Potato field, Hamadan	1 / 60	1 / 240
Wheat field, Hamadan	0 / 20	0 / 123
Wheat and Barley field, Lorestan	0 / 31	0 / 711
Summer crops and Vegetable field, Lorestan	0 / 27	0 / 535
Wheat field, Khuzestan	0 / 49	0 / 626
Vegetable field, Khuzestan	0 / 44	0 / 974
Sugar beet field, Zanjan	0 / 30	0 / 100
Wheat field, Zanjan	0 / 50	0 / 159
Wheat field, Qazvin	1 / 12	1 / 103
Sugar beet field, Qazvin	1 / 16	1 / 185
Wheat field, Azarbaijan Gharbi	0 / 8	0 / 282
Sugar beet field, Azarbaijan Gharbi	3 / 23	4 / 604
Summer crops and Vegetable field, Azarbaijan Gharbi	1 / 16	1 / 376

Table 1. Distribution of *Bacillus thuringiensis* (*Bt*) in soil samples collected from Agricultural fields of different provinces of Iran and frequency of crystal-forming isolates in bacterial colonies examined (Isolates/Colonies) (Continued)

Type of cultivation and location	Frequency of *Bt* in samples examined	Isolates/Colonies
Sunflower field, Azarbaijan Gharbi	3 / 10	3 / 298
Wheat field, Azarbaijan Sharghi	0 / 7	0 / 235
Sugar beet field, Azarbaijan Sharghi	0 / 7	0 / 238
Potato field, Azarbaijan Sharghi	1 / 5	2 / 156
Vegetable field, Azarbaijan Sharghi	1 / 12	1 / 393
Potato field, Ardabil	2 / 9	3 / 255
Wheat field, Ardabil	5 / 22	5 / 735
Cotton field, Ardabil	3 / 6	6 / 76
Rice field, Gilan	17 / 243	19 / 2442
Rice field, Mazandaran	2 / 60	3 / 747
Cotton field, Golstan	3 / 13	4 / 331
Wheat field, Golstan	8 / 35	8 / 1094
Summer crops and Vegetable field, Golstan	0 / 54	0 / 989
Cotton field, Khorasan	0 / 20	0 / 347
Sugar beet field, Khorasan	0 / 12	0 / 151
Wheat and Barley field, Khorasan	0 / 20	0 / 263
Wheat field, Sistan and Baluchistan	1 / 36	1 / 386
Summer crops and Vegetable field, Sistan and Baluchistan	0 / 46	0 / 495
Vegetable field, Hormozgan	2 / 72	3 / 932
Vegetable field, Bushehr	3 / 17	3 / 227
Sesam field, Bushehr	0 / 21	0 / 217
Oil seed rape field, Bushehr	4 / 44	4 / 583
Cotton field, Fars	0 / 20	0 / 299
Wheat field, Fars	0 / 55	0 / 780
Corn field, Fars	1 / 27	1 / 450
Wheat and Barley field, Kerman	2 / 30	2 / 552
Summer crops field, Kerman	2 / 72	2 / 1031
Vegetable field, Yazd	0 / 10	0 / 134
Wheat and Barley field, Yazd	4 / 68	5 / 349
Wheat and Barley field, Esfahan	0 / 33	0 / 391
Potato field, Esfahan	0 / 37	0 / 553
Summer crops and Vegetable field, Esfahan	1 / 24	1 / 305
Sunflower field, Esfahan	4 / 36	4 / 318
Wheat field, Tehran	1 / 32	1 / 329
Potato field, Tehran	0 / 5	0 / 107
Vegetable field, Tehran	0 / 4	0 / 135
Summer crops, Semnan	0 / 30	0 / 204
Wheat field, Markazi	2 / 38	2 / 446
Summer crops and Vegetable field, Kohkiluyeh and Bovirahmad	0 / 39	0 / 351
Sugar beet field, Kohkiluyeh and Bovirahmad	5 / 29	7 / 500
Wheat field, Chaharmahal and Bkhtiari	8 / 72	10 / 867
Vegetable field, Chaharmahal and Bakhtiari	1 / 29	1 / 174
Vegetable field, Qom	2 / 40	2 / 120

Table 2 shows the results of the pathogenicity of different morphological types of parasporal inclusions against the cotton bollworm (Order Lepidoptera) and the Colorado potato beetle (Order Coleoptera). Our study on the toxicity of different isolates revealed that fifty-eight % of isolates were pathogenic to cotton bollworm, *H. armigera* and Colorado potato beetle, *L. decemlineata*, leading to more than 10% mortality. The results also showed that about 70% of the effective isolates had a bipyramidal parasporal inclusion. Furthermore, about 15.6% were cuboidal, 6.3% irregular, 2.3% spherical and 7.8% isolates had both bipyramidal and cuboidal inclusion.

Table 2. Pathogenicity of different morphological types of parasporal inclusions against the cotton bollworm (CBW: *Helicoverpa armigera*) and the Colorado potato beetle (CPB: *Leptinotarsa decemlineata*)

Insects tested	Percentage of active[a] isolates in each morphological group[b]					Number of isolates tested
	B	C	S	I	BC	
CBW	72.9	8.3	4.2	10.4	4.2	48
CPB	45	25	0	5	25	20
CBW + CPB	85.7	14.3	0	0	0	7

[a] More than 10% corrected mortality.

[b] Classification of morphological types of parasporal inclusions: B, bipyramidal; C, cuboidal; S, spherical; I, irregular; BC, bipyramidal and cuboidal

About 36.7% of isolates in which most of them (72.9%) had bipyramidal prasporal inclusion were pathogenecic on cotton bollworm and 14.1% of isolates in which most of them (45%) had bipyramidal inclusion were pathogenic on Colorado potato beetle. 7.8% of isolates were pathogenic on both insects, in which most of them (85.7%) had both bipyramidal and cuboidal parasporal inclusion. 3.9% of the isolates caused 70-100% mortality on cotton bollworm and 1.56% on Colorado potato beetle after 3 days. There is no direct correlation between source of *B. thuringiensis* isolates and pathogenicity.

There is no clear relation between the pathogenecity and parasporal inclusion shape except that isolates harboring spherical parasporal inclusions were effective on cotton bollworm. Bernard *et al.* (1997); Martin and Travers (1989) and Ohba and Aizawa (1986) got the same results regarding the pathogenecity and parasporal inclusion shape. Some researchers already reported that non-pathogenic *B. thuringiensis* are common in soil (Martin and Travers, 1989; Ohba and Aizawa 1986); our experiments also testify these results.

The ecological survey of *B. thuringiensis* has shown that soil is the habitat for this bacteria and different isolates with different phenotypes and genotypes are strongly related to the geographical distribution of the soils.

REFERENCES

A. J. De Lucca, J. G. Simonson, and A. D. Larson,. *Bacillus thuringiensis* distribution in soils of the United States. *Can. J. Microbiol.* 27, 865-870 (1981).

Anwar Hossain, , M. Sohel Ahmad and Sirajal Haque, Abundance and distribution of *Bacillus thuringiensis* in the agricultural soil of Bangladesh. *J. Invertebr. Pathol.* 70, 221-225 (1997).

C. N. Chilcott, and P. J. Wigley, Isolation and toxicity of *Bacillus thuringiensis* from soil and insect habitats in New Zealand. *J. Invertebr. Pathol.* 61, 244-247 (1993).

H. T. Dulmage, and K. Aizawa, Distribution of *Bacillus thuringiensis* in nature. *In* "Microbial and viral pesticides". (E. Kurstak Ed.), 209-237 (1982).

K. Bernhard, P. Jarrett, M. Meadows, J. Butt, D. J. Ellis, G. M. Robert, S. Pauli, P. Rodgers, And H. D. Burges, Natural isolates of *Bacillus thuringiensis*. Worldwide distribution, characterization and activity against insect pests. *J. Invertebr. Pathol.* 70, 59-68 (1997).

L. E. Padua, B. P. Gabriel, K. Aizawa, And M. Ohba, *Bacillus thuringiensis* isolated from the Philippines. *Philipp. Entomol.* 5, 199-208 (1982).

M. Ohba, and K. Aizawa, Distribution of *Bacillus thuringiensis* in soils of Japan. *J. Invertebr. Pathol.* 47, 277-282 (1986).

P. A. W. Martin, and R. S. Travers, Worldwide abundance and distribution of *Bacillus thuringiensis* isolates. *Appl. Environ. Microbiol.* 55, 2437-2442 (1989).

R. Marzban, Comparative bioassay of some native isolates of *Bacillus thuringiensis* and serotype of *kurstaki* on Indian meal moth (*Plodia interpunctella* Hb.). *Appl.Entomol. and Phytopathol.* 70, 29-36 (2002).

S. H. Kim, D. Lee, S. D. Woo, Y. M. Yu, and S. K. Kang, Distribution, Serological identification and PCR analysis of *Bacillus thuringiensis* isolated from soils of Korea. *Cur. Microbiol.* 37, 195-200 (1998).

T. Attathom, W. Chongrattannameteekul, J. Champaisang, and R. Siriyan, Morphological diversity and toxicity of delta - endotoxin produced by various isolates of *Bacillus thuringiensis*. *Bull. Entomol. Res.* 85, 167-173 (1995).

W. S. Abbott, A method of computing the effectiveness of an insecticide. *J. Econ. Entomol.* 18, 265 – 267 (1925).

In: Biotechnology, Agriculture and the Food Industry
Editor: G. E. Zaikov, pp. 101-104 ISBN 1-60021-040-6

Chapter 17

SEARCH FOR T-DNA-LIKE SEQUENCES IN PLANT GENOMES, USING REAL-TIME PCR WITH DEGENERATE PRIMERS AND PROBE

T. V. Matveeva*[*], *D. I. Bogomaz and L. A. Lutova
St.Petersburg State University,
7/9, University emb., 199034 St.Petersburg, Russia

ABSTRACT

Taq Man modification of real time PCR with degenerate primers and probe was examined as method for analysis of plant species on presence of oncogene-like sequences. The results of analysis were compared with ones obtained by traditional for this purpose methods. Real time PCR with degenerate primers and probe has shown itself as quick and adequate method and its results do not contradict to ones, obtained by traditional methods.

Key words: *Agrobacterium, plants, horizontal gene transfer, Real time PCR*

INTRODUCTION

Problem of horizontal gene transfer from Procariota to Eucariota is one of the most intriguing. A number of examples of T-DNA like sequences in some representatives of Nicotiana genera support the hypothesis about horizontal gene transfer in nature. Search for new examples of horizontal gene transfer is an actual problem, because description of oncogenes in agricultural species, used for thousands years, would decrease public concern about safety of transgenic food. Presence of T-DNA like sequences in wild species would decrease public concern about their ecological safety.

[*] St.Petersburg State University, 7/9, University emb., 199034 St.Petersburg, Russia. E-mail: radishlet@yahoo.com, fax: +7(812)4287733

Until now research work in this field was carried out using such methods as:

- Southern blot hybridization of genomic DNA, followed by cloning and sequencing of fragments, demonstrating positive hybridization signal [1,2];
- classical variants of PCR followed by cloning and sequencing of fragments, demonstrating positive hybridization signal [3]

Development and wide spread of Real time PCR lets to apply it for this task. TaqMan modification of Real-time PCR lets to combine positive sides of PCR and Southern hybridization in single reaction. We proposed, that usage of degenerate primers and probe, corresponding to the concervative parts of gene, would lead to detect sequences, which differ from ones, described earlier.

The aim of this work is to characterize express method for search for new examples of horizontal gene transfer from agrobacteria to plants.

The particular tasks were:

- to perform Real-time PCR with degenerate primers and probe to genes rolB, rolC, ORF13, ORF14 on DNA of species, previously analyzed on presence of these oncogenes by traditional methods [1-4];
- to compare results, obtained by real time PCR, with results, obtained by traditional methods.

Material and Metods

Following plant species were used as material: Nicotiana glauca, Nicotiana tabacum, Nicotiana alata, Nicotiana rustica, Nicotiana langsdorffii, Nicotiana longiflora, Petunia hybrida, Cestrum parqui, Solanum tuberosum, Lycopersicon esculentum, Capsicum annum, Cucumis sativus, Brassica oleraceae, Raphanus sativus, Arabidopsis thaliana, Betula pendula, Daucus carrota, Zeya mais, Triticale.

DNA was isolated by method , described by Dellaporta [8].

With the aim to check DNA quality we carried out control PCR with random primers A1 и A2 (Roth) inreaction volume of 50 ul, using “Tercyc” gene cycler from («DNA Technology», Russia). Reaxion mixture contained 100ng of DNA, 10pM of primer, 2.5 U of Taq polymerase (Sileks M), enzyme buffer provided by this company, 200uM of each dNTP. PCR was carried out according the following programme: 5 min. - 93°C, 33 cycles (17 sec. - 93°C, 40 sec. - 37°C, 40 sec 72°C), followed by 5 min at 72°C.

Fragments were separated in 1% agarose gel on 1x TAE buffer. 100bp+1.5 kb ladder («Sibenzyme») was used as molecular weight marker.

For Real time PCR we used 22-24-mer primers to conservative parts of four oncogenes of *Agrobacterium rhizogenes*:

RolCL: gtcgactgcccgacgatgatgc;
RolCR: ggcgaagtaggcgctccggata;
RolBL: atgcagaaagtgctggaggaa;

RolBR: gtgcggccaagcaaggttgtg;
ORF13L: ctttacctccmwggccgtggc;
ORF13R: ccagtgagacatttgtctgat;
ORF14L: gstkgaggcggmgctackdgacta;
ORF14R: atcaacaatggagttgagtttctc;

and following degenerate probes to the conservative parts of corresponding genes:

RolB_probe: R6G-atcgtbgtcgcagcacggcctccattcca-BHQ1;
RolC_probe: Cy5-tcmtcaccaatcttccgccgtaccag-BHQ2;
ORF13_probe: Rox-aggaacgacgcasgmtaacagntccccatc-BHQ2;
ORF14_probe: FAM-ctcgatcarwctgtgackgtcgacaagc-BHQ1.

Primers and probes are synthesized by Syntol company.

The reaction volume was 30 ul and contained 100ng of DNA, 1,5U of Taq polymerase (Sileks M), enzyme buffer provided by this company, 200uM of each dNTP, 250 pM of each primer and 250-500 pM of probe (depending on number of degenerate position and type of fluorescent dye).

PCR was carried out according the following programme: 5 min. - 93 °C; 40 cycles at 15 sec. - 93 °C, 60 sec. - 45 °C, 60 sec. - 72 °C, followed by 15 sec - 25 °C , using POP 16 gene cycler (Instituter for Analytical Instrumentation of Russian Academy of science).

Specific PCR products were isolated from the gel, filtered through the glasswool ("Sigma"), ethanol precipitated, dissolved in water and sequenced by "Omnix" company.

RESULTS AND DISCUSSION

After examination of DNA samples quality by RAPD PCR we performed Real-time PCR with degenerate primers and probe on these samples. Successful Real-time PCR reaction, where specific product was amplified, was detected by growth of fluorescence, which mathematically transforms in exponential curve.

Such picture was observed in the experiment with species, where sequences, homologous to T-DNA were shown before: *Nicotiana glauca, Nicotiana tabacum*. Sequencing of PCR products with all used primers on the termplate DNA of N.glauca was done. Comparing of the results of sequencing with databases from Interner has shown their identity to the corresponding *N.glauca* genes and similarity to agrobacterial oncogenes. So our data completely correspond to data obtained before for these two species.

Using this method following species were analyzed Nicotiana alata, Nicotiana rustica, Nicotiana langsdorffii, Nicotiana longiflora, Petunia hybrida , Cestrum parqui, Solanum tuberosum, Lycopersicon esculentum, Capsicum annum, Cucumis sativus,Brassica oleraceae, Raphanus sativus, Arabidopsis thaliana, Betula pendula, Daucus carrota, Zeya mais, Triticale. They show absence of growth of fluorescence in real time PCR reaction with primers and probes to the oncogenes rolB, rolC, ORF13, ORF14 from A. rhizogenes. These data do not contradict with data, obtained by traditional methods for these species[1-4].

Summarizing our results, we can say, that Taq Man Real-time PCR with degenerate primers and probe is adequate method for the purpose of search for T-DNA-like sequences in plant genomes.

By this method it is possible to analyze big numbers of samples for the short period of time. In this case limiting stade is plant DNA isolation. Besides method lets to carry several reaction in single reaction mixture. In case of our PCR machine, made in the Institute for Analytical Instrumentation of Russian Academy of Sciences, using primers and probes from "Syntol" this number is four. All these facts let to make a conclusion, that this method can be applied for screening of big numbers of species.

The research described in this publication was made possible in part by Award No.ST-012-0 of the U.S. Civilian Research and Development Foundation for the Independent States of the Former Soviet Union (CRDF), Russian Foundation for Basic Research (project no. 02-04-48800)

REFERENCES

[1] R.E. White, et al.: *Nature,* 301(5898), 348-350(1983).

[2] A.D. Meyer, T. Ichikawa, F .Meins: *Mol. Gen. Genet.,* 249, 265-273 (1995).

[3] M.C. Intrieri, M. Buiatti: *Molecular Phylogenetics and Evolution*, 20(1), 100-110 (2001).

[4] T.V. Matveeva, L.A. Lutova, D. Wood, E.W.: *Biology of Plant-Microbe Interactions*, 4, 526-529 (2003).

[5] S.L. Dellaporta et al.:Molecular Biology of Plants: *A laboratory Course Manual*. 36-37(1983).

In: Biotechnology, Agriculture and the Food Industry
Editor: G. E. Zaikov, pp. 105-110
ISBN 1-60021-040-6

Chapter 18

On the Conception of Creating the International Cryobank and the Use of Epibrassinolid to Increase Survival of Gametal Cells of Valuable Fish Species in Crioconservation Process

Mikhail A. Egorov [*]
Astrakhan State University, Astrakhan, Russia

Abstract

Astrakhan State University within the limits of the laboratory of biotechnologies is involved in research work aimed at preservation of the gene pool of rare and threatened species of fish (sturgeon and salmon), plants and animals of the Volgo-Caspian basin. The main purpose of the project is the creation and realization of effective work of the international Caspian cryobank, actual recovery of biological resourses of the Caspian region in all genetic and species diversity, which will make a significant scientific, practical and economical contribution into the development of the industries of the Caspian Sea countries. The research work is sponsored by the grant of the President of the Russian Federation aimed to support young Russian scientists and the leading scientific schools of the Russian Federation.

Keywords: *Cryobank; Cryoconservation; Epibrassinolid; Valuable biological objects*

Introduction

The conditions of the historical development of the Caspian Sea created a unique reservoir which has no analogues in the world. According to the existing circumstances,

[*] Doctor of Biology, Head of Biotechnologies Chair, Head of Biotechnologies Laboratory. *Tatischev str. 20a, Astrakhan, 414056, Russia.* Phone: (8512) 25-17-54; Fax: (8512) 25-17-18; E-mail egorov@astranet.ru

northern and tropical spicies of flora and fauna successfully live together [1, 2, 3]. So along with the material importance, this reservoir has a great scientific and economical importance. As a result of human activities the ecological situation in this region is difficult and becomes worse year after year. Uncontrolled sea fishery and the growing anthropogenic activity led the ecosystems of the Caspian Sea out of balance and put them on the verge of inevitable degradation [4-6]. The evidence of this is the quantitative and qualitative changes in the forage reserves of the finishing areas of the northern part of the Caspian Sea, the decrease of the number of all species of sturgeon and the rejuvenation of their shoals [3, 5, 7]. Other representatives of water life are not in better conditions. The disappearance of traditional spawning places on the biggest rivers and the shortening of the forage reserves which followed threatens the life of almost any species of fish in the Caspian Sea [1, 6]. It is obvious that the disappearance of any single species of fish, even if it is not valuable from the point of view of industries, will produce a chain reaction of the downfall of the flora and fauna of the region [8, 9, 10]. Such a prospect threatens not only Russia, but it is an international problem which touches upon the interests of all countries situated on the shore of the Caspian Sea.

The way out from the present situation is in the conception of academician Pavlov D. which provides seven independent but at the same time interrelated approaches, among which the most promising is genotypic. This approach is based on the fundamental scientific statement that al the hereditary information about the features and characteristics of the species is stored in genes, the totality of which presents genotypes of individual organisms [13].

The genotypic principle as a separate method presupposes securing of a long storage of genotypes – the creation of genetic banks of rare and threatened species. It is most important where the reserves for preservation of natural populations of a species are exhausted, and also where the uncontrolled introduction and hybridization leads to the loss of pure natural populations, to the loss of the gene pond [8-15].

2. Brief of Previous Work

The principle scheme for creating of cryobanks and restoration of organisms from the cryoconservated material is worked out by professor Venprintsev B. in the late 70-s of the XX century in Russia [7]. The working out of the methods of restoration of the living organisms is done in different countries, and they give encouraging results. At present, several genetic cryobanks function in Russia with sufficient degree of reliability, including those situated in the village Ribnoe in the Dmitrovsky district of the Moscow oblast and in the town Puschino-na-Oke, whose actions are aimed at creating of collections of gene pool of the animals, aquatic life and plants and at the solution of scientific problems in this field [10, 11, 14, 15, 16]. The creation of a new cryobank will be of international importance, and will undoubtfully stimulate the economical development of sectors of the national economy of the countries near the Caspian Sea [10,15].

The question of the protection and conservation of durable and high-quality sperm - the bearer of genetic information and the variety of species of valuable species of sturgeon – is a topical on nowadays. One of the promising solutions of this problem is the cryoconservation of sperm, the creation of cryoconservated collection of cells of valuable species of fish, the creation of a cryobank. Cryoconservation of spermatozoa is the most developed field of cryobiotechnology [8,10,17,18-21]. In Russia low-temperature conservation of gametes has been used for several years in the form of methods proved at the sperm of carps, sturgeon and salmon species of fish [8]. During these years the necessity of the working out and creation of an international cryobank of genetic material of the species of aquatic life of the Caspian Sea, especially of sturgeon, has become more apparent [15].

3. Results

Astrakhan State University within the limits of the laboratory of biotechnologies is carrying out an intensive research work aimed at preservation of the gene pool of rare and threatened species of fish (sturgeon and salmon), plants and animals of the Volga-Caspian basin [10, 15]. Having efficient specialists (the group has been working at this problem since 1994) in the field of low-temperature genetic banks and having necessary premises, Astrakhan State University is a real basis which can be used for creating an international Caspian cryobank [15]. The research work is sponsored by the grant of the President of the Russian Federation aimed to support young Russian scientists and the leading scientific schools of the Russian Federation given for the term of three years within the limits of the research work entitled "The organization of the Low-Volga regional cryobank of the gene pool of the flora and fauna of the Volga-Caspian basin".

The project is realized in three phases.

On the first stage is the organization of a cryostorage of limited capacity.

Large scope investigations were carried out to estimate crioprotective qualities of bioactive substance (BAS), epibrassinolid. Highly effective epibassinolid has been developed by NNPP «NEST M» Moscow and is produced as Epin Extra preparation. In our work it was used to increase the activity of spermatozoids of valuable fish species after crioconservation. The main investigations objects were males of white sturgeon (Huso Huso L.) and sheefish (Stenodus leucichthys leucichthys G.). Crioconservation and unfreezing as well were carried out in accordance with the following methodics [7, 9]. After the samples were unfrozen we determined the percentage of alive cells and the period of mobility while activating with water and used BAS solutions. The most optimal concentrations of Epin-Extra were determined to be 10^{-12} and 10^{-14} mg/l [15]. The present preparation is turned out to be more effective in comparison with earlier existed preparations on the base of epibrassinolid (Egorov, Vivitskaya, 1998). After crioconservation the period of cells` life was 2 less (45,6%) than for native samples (92%). The use of preparations on the base of epibrassinolid NNPP «NEST M» as activating environment increased the period of cells` life, after crioconservation in average 1,6 (71,7 of the initial). The general data on crioconservation of fish cell with the use of effective preparations NNPP «NEST M» point that this kind of investigation is very perspective. The purpose of this cryostorage is to provide the basis for wide-range research work aimed at working through the modes of freezing and defreezing of the examples of different species of biological objects, at the selection of cryoprotectors for traditional and new species, at searching for the ways of freezing of the gametes of the species of aquatic life, flora and fauna of the region without damaging them.

On the second stage – the creation of material and technical basis of the cryobank, working through all the phases in commercial prototype mode.

The third stage presupposes the settling of the cryobank in operation and bringing it to its planned production capacity, and accumulation of the genetic material for scientific, industrial and commercial purposes and their realization.

The primary aim of this project is organization and putting into practice of effective functioning of international Caspian cryobank. Actual restoration of biological resources of the Caspian region in their genetic and species variety, which will make an important contribution into scientific and technological development of a number of industries of the countries situated near the Caspian Sea.

The creation of an international Caspian cryobank will allow:

1) to preserve the genetic information of rare and threatened species of fish, plants and of the best breeds of animals of the Volga-Caspian basin at temperature of liquid nitrogen for decades without loosing the genetic standard.
2) to transport (to negotiate) the genetic material to the districts where threatened species or species whose population is quickly reducing are found in order to restore their population or to sell the genetic material in accordance with the demands of the customer.
3) to provide an opportunity for selective and genetic works in fishing and agricultural enterprises of the region.
4) to correct constantly "the influence of fresh blood" in the technology of sturgeon-fishing plants of the region as the amount of the producers of rare and threatened species of fish (lamprey, sturgeon, white sturgeon, etc.), of plants (the endemics of the region) and of the best breeds of farm animals.
5) to create a full genetic collection of different species of aquatic life for the restoration of ichtyofauna of the basin and ecosystems, which is possible due to the almost unlimited capacity of the storage pools.
6) to carry out wide-range research work in promising fields of modern biology and aimed at development of the agricultural industry of the countries close to the Caspian Sea.

The methods of cryoconservation which we are working out are quite simple, they are provided with necessary equipment and are available for the organization of an up-to-date cryobank and for the wide practical usage of biological materials of a cryostorage at the enterprises of pre-Caspian countries. These methods have a great scientific and economical importance for the development of agricultural industries of the countries which participate in the project.

For realization of this scientific project Astrakhan State University has an experimental complex of laboratories of biotechnologies with special rooms for cryobank (Astrakhan oblast, Privolzhskiy district, village Nachalo, scientific housing of the agricultural faculty of Astrakhan State University); c. Astrakhan, the building of the Institute of Natural Science of the Astrakhan State University; Collaborators of the laboratory of biotechnologies also carry out their work at the enterprises of corresponding types: Kizansky sturgeon fishbreeding plant (Astrakhan oblast, Privolzhsty region, village Kizan); Aleksandrovsky sturgeon fishbreeding plant (Astrakhan oblast, Ikryaninsry region, village Trud-Front) and at some others.

Our researches are carried out in cooperation with a number of the Russia's biggest scirntific institutions: Interdepartmental Ichtiological Commission (Moscow), The Institute of Biophysics of a Cell (Puschino-na-Oke), All-Russian research institute of pond fish industry (Dmitrov, Moscow oblast), etc.

4. CONCLUSION

Preliminary calculations showed that at present fishbreeding plants and agricultural enterprises of pre-Caspian countries are in need of genetic material. The demand on the genetic materials of the European countries were not counted [4, 8, 10, 15].

So the creation of an international cryobank of the gene pool of the flora and fauna of the Caspian basin on the basis of Astrakhan State University will make an important scientific and economical contribution into the development of the industries of national economy of pre-Caspian countries. The important aspects of functioning of the cryobank are: considerable improvement and further modern development of the selective activities; selection and preservation of the genetic material of the most valuable representatives of the flora and fauna of the region; searching and obtaining of valuable genetic material in Russia and abroad for further improvements; increasing the crop capacity of the main agricultural plants in the region and increasing of the survival rate of farm animals and valuable species of fish. Replenishment of natural populations (pure lines) of food fish by the genetically high-quality fish, grown at the enterprises for artificial reproduction of valuable species of animals in Russia, Iran, Kazakhstan and other pre-Caspian countries.

REFERENCES

[1] Ivanov V.P. The condition and the prospects of raw-material sources in the Caspian Sea //*The first congress of Russian ichtiologists. Theses of the reports.* Moscow. 1997 p. 421. (in Russian).

[2] Malutin V.S. The tendencies of the development of Russian sturgeon-breeding // *Informational packet Fish Industry.* Aquaculture series. Sturgeon is the promising object of aquaculture. Moscow. 1994. №1. p. 6-10. (in Russian).

[3] Khodorevskaya R.P. The calculation of the stores of sturgeon in the Cspian Sea // *Fish Industry* №5. 1997. p. 39 (in Russian).

[4] Vitivitskaya L.V., Tikhomorov A.M., Egorov M.A. *Sturgeon of the world ocean.* Astrakhan, Astrakhan Technical State University. 2002. p. 160 (in Russian).

[5] Zhuravleva G.F., Zemkov G.V. Structural abnormalities in ovocytes depending on pathomorphogenesis of the sturgeon liver during the cumulative toxicosis // *Theses of an international scientific conference "Sturgeon in the XXI century"* Astrakhan, 2000, p. 145. (in Russian).

[6] Ivanov V.P., Vlasenko. A.D., Khodorevskaya R.P. Ways of preservation of sturgeon // *Fish Industry.* №2. 1995, p. 24-26. (in Russian).

[7] Egorov M.A., Vitvitskaya L.V. The usage of bioactive substances in artificial reproduction of sturgeon in the Volga-Caspian region. Astrakhan, ATSU, 2002, p 100. (in Russian).

[8] Ananiev V.I. Conception of preservation and stable usage of bio-diversity using the methods of cryoconservation of genomes of species of aquatic life // Fish Industry. Aquaculture series.1997 №1, p. 1-36. (in Russian).

[9] Egorov M.A. *Physiological peculiarities of the influence of phytohormone of epibrassinolide on the organism of animals in early ontogenesis*. Astrakhan. ASPU, 2002, p. 269. (in Russian).

[10] Egorov M.A. The organization of a cryobank of the gene pool of valuable flora and fauna of the south of Russia // "*Modern Science intensive technologies*". №1. M. 2004, p. 54-55. (in Russian).

[11] Ananiev V.I.,Tsvetkova L.I., Manokhina M.S. Problems of preservation of the genomes of salmon and sturgeon // *Fish Industry*. Aquaculture series. 1998. №1, p 2-24. (in Russian).

[12] Lukianenko V.I. *General Ichtiotoxicology*. M. 1983, p.320. (in Russian).

[13] Pavlov D.S. Approaches to the protection of rare and threatened species of fish // *Materials to the 38 plenum of the Ichtiological Commission*. M. 1990, p 12-13. (in Russian).

[14] Anaiev V.I., Andreev A.A., Gokovanova T.S., Petropavlov N.N., Tsvetkova L.I. Experience in cryoconservation of sperm of white sturgeon // *Fish Industry. Aquaculture series*. 1998. №1, p 25-32. (in Russian).

[15] Egorov M.A. About the creation of an international cryobank of the gene pool of the flora and fauna of the Caspian basin // *m. Cytology*. v. 46. № 9. St. Petersburg, 2004, p. 791-792. (in Russian).

[16] Karnaukhov A.V. To the question of stability of the chemical composition of the atmosphere and the thermal balance of the Earth // *Biophysics*, v. 39, №1, Puschino, 1994, p. 148-152. (in Russian).

[17] Tsvetkova L.I., Savushkina S.I., Titareva L.N., Dokina O.B., Pronina N.D. *Methodical Aids for cryoconservation of the sperm of carp, salmon and sturgeon*. M. 1997, p.11. (in Russian).

[18] Ananev V.I. Problems of conservation and stable employment of biological diversity with the use of deep freezing methods for hydrobionts /*Froid et Aquaculture. Refrigeration and Aquaculture*. Bordeaux, France. - 1996. - P. 33-42. (in Russian).

[19] Egorov M.A. The secret land //*International Journal «Caspian Research»* Book. N. 5. Baku, 2003. P. 85 – 88. (in English).

[20] Egorov M.A., Vitvitskaya L.V. The evaluation of biologically active substances influence on sturgeon embriogenesis under unfavorable conditions //*The Caspian sea science education technologies. International journal of collected academic articles*. Book I. Astrakhan. 1998. P. 104-109. (in English).

[21] Wildt D.E., Rall W.F., Seal U.S. Establishing genetic resource banks for endangered species //*Nature*, Vol.167, 1990, - P. 342-383. (in English).

In: Biotechnology, Agriculture and the Food Industry
Editor: G. E. Zaikov, pp. 111-118
ISBN 1-60021-040-6

Chapter 19

STRAIN *BACILLUS SUBTILIS 8130* - THE NATURAL STIMULATOR OF DIGESTION

*N. A. Ushakova**, *A. A. Kozlova and A. V. Nifatov*
Severtsov Institute of Ecology and Evolution of the Russian Academy of Science, Moscow

ABSTRACT

Strain *Bacillus subtilis 8130* is allocated from blind gut wild herbivorous birds (wood grouse). It is the natural catalyst of processes of digestion of vegetative components of forages due to formation of enzyme endoglucanase and allocation in cultural liquid metabolites, stimulating development of lactobacillus, lactic streptococci. Participation in digestion of cellulose is confirmed with improvement zootechnik parameters of experimental animals at introduction cultural liquid by bacilli into a beer pellet and the subsequent feeding pellets with probiotic an animal.

Key words: *probiotic, endoglucanase, metabolites.*

Bacteria of species *Bacillus subtilis* are frequently used in probiotic preparations. It is considered, that their influence on an organism of human and animals is mainly connected to processes of immunomodulation [1,2] and the ability to synthesize antibiotics. However rich metabolic opportunities and adaptable plasticity of the given bacteria allow finding out new effects. One of properties *B.subtilis* is participation during digestion of vegetative components of forages.

B.subtilis is widespread species of microorganisms. The vegetative rests are a habitat of its dwelling. With vegetative forages *B.subtilis* gets in a gastro enteric path of animals, but it is not considered resident intestinal symbiotic bacteria, and its role in digestion is not considered as essential. Meanwhile it is obviously important the fact of allocation *Bacillus* from intestines of various kinds herbivorous animals - rodents, hares, termites [3,4,5].

* Severtsov Institute of Ecology and Evolution of the Russian Academy of Science 119071, Moscow, Leninsky prospect, 33. E-mail: ushakova@sevin.ru, fax +7 (095) 954-55-34

From a blind gut wood grouse during the winter period when the bird ate mainly needles, it is allocated strain *B.subtilis 8130* [6]. This strain is the natural catalyst of processes of digestion of organic substances, and to the greatest degree - cellulose [7].

The purpose of this work is studying of mechanisms of probiotic actions *B.subtilis 8130*: abilities of bacteria to produce endoglucanase and metabolites, stimulating development of ruminococcus, lactobacillus, lactic streptococci.

METHODS

The strain *L.acidophilus Suav. B-4625* used in this work and from the intestines of a rabbit - *Streptococcus lactis*. Identification of a streptococcus carried out with the help [8]. About nonpathogenic properties of the allocated bacteria judged on absence of rejections from physiological norm at white mice in weight 18 - 20 g and pathological changes in internal organs of these animals at peritoneal introduction on 0.5 ml of suspension of cells of two-daily cultures in a physiological solution with a credit $1 \cdot 10^9$ cells / ml and the subsequent supervision over animals within 15 day.

Endoglucanase activity determined by viscosimetric method [9], using capillary glass viscosimetr with a capillary in diameter of 0.73 mm, carboxymethylcellulose sodium salt average viscosity firms Serva (USA), the phosphate buffer, pH 6.86. Initial solution carboxymethylcellulose sodium salt of concentration 15 g/l prepared by dissolution such salt in the buffer at 40-60^{0}C. Necessary amount initial carboxymethylcellulose sodium salt solution diluted before the beginning of experience with the phosphate buffer so that time of its expiration made 110 - 130 sec. Cultural liquid separated from cells by centrifugation (15 min, 1500g), and the activity of endoglucanase enzyme complex defined in supernatant by viscosimetric method. Originally in viscosimetr poured 10 ml of the buffer and measured time of expiration Tб. Then the buffer merged and poured 10 ml of solution carboxymethylcellulose sodium salt, measured time of expiration T_0. After that in viscosimetr added 1 ml of the supernatant and marked on a stop watch time of the beginning of reaction of reduction of viscosity of solution carboxymethylcellulose sodium salt. Time of the expiration of reactionary mix Tt measured every 5 minutes during 30 minutes with the subsequent counting in the program of mathematical modeling (MathCAD). The unit under the formula expected activity:

$$A = \frac{K \cdot (T_0 - T_t)}{T_б \cdot \left(\frac{T_0}{T_б}\right)^{7/8} \cdot \frac{V_{enzyme}}{V_{substrate}} \cdot (T_{experiment} + \frac{T_t}{60})} = \left[\frac{\mu mol}{\min \cdot ml}\right]$$

Where Tб - time of the expiration of the buffer, T_0 - time of the expiration of solution carboxymethylcellulose sodium salt, Texperiment = 5 mines (or 10-15-20-25-30 min), Tt - time of the expiration of a reactionary mix, K - a constant, for carboxymethylcellulose sodium salt average viscosity value 10 is accepted. All "T" except for T experiment (which expressed in minutes) expressed in seconds. V enzyme = 1ml, V substrate = 10 ml.

Definition of stimulating effect carried out by the developed original technique. In test tubes in volume of 50 ml poured on 20.0 ml sterile glucose-peptone nutrient medium, g/l: glucose - 20.0; peptone - 4.0; a barmy extract - 1.5; $(NH_4)_2HPO_4$ - 1.5; $MgSO_4$ - 0.5; pH - 6.9, and also brought on 0.5 ml of suspension of corresponding test - culture of bacteria. Then in control test tubes added on 0.5 ml of a sterile nutrient medium for development of *B.subtilis 8130*. In skilled variants into test tubes entered 0.5 ml of the cultural liquid of *Bacillus subtilis 8130*, brought up on the same nutrient medium. From each test tube after hashing selected aliquot in which defined initial amount of cells of tests - cultures and *B.subtilis 8130*. Further test tubes cultivated at 30°C within 1-3 day, selecting through each 24 hours of test. The account of number of cells carried out by direct their calculation in fixed smears, painted gentianviolet, in sight microscope Axioskop (firms Zeiss, Germany) on method [10]. Statistical data processing and construction of schedules carried out, using program Origin 4.

Index of stimulation defined under the relation of amount of cells of daily tests - cultures in 1 ml the glucose-peptone nutrient medium at presence *B.subtilis 8130* to number of cells in 1 ml of daily culture in the control without *Bacillus*.

Cultivating *of Bacillus subtilis 8130* carried out on the glucose-peptone nutrient medium, on the nutrient broth, and also on nutrient mediums containing a corn flour or corn gluten, a soya flour, sunflower presscake, beet bagasse.

Experiences on feeding dwarfish rabbits carried out on Chernogolovka scientific - experimental base of Institute. Groups of animals made on 10 individuals in the age of 12-14 weeks. Results shorthanded with use of program Statistica 6.0. Authors express gratitude doct. Kotenkova E.V. for work under the maintenance and feeding of rabbits.

Experimental feeding calf was carried out by employees of laboratory of feeding of the All-Russia institute of animal industries of Russian Academy of Agrarian Sciences. Authors express gratitude prof. Kirilov M.P. and to his colleagues for carrying out of experiment.

DNA fingerprint was carried out by Center of Bioengineering, Russian Academy of Sciences. Authors express gratitude doct. Kuznetsov B.B.for testing strains of *Bacillus*.

Results

Cells *B.subtilis 8130* are capable of forming poorly absorbed endogluconase. The size of the enzyme activity grew after the first day of development of culture (fig. 1, table 1). The enzyme complex was constitutional since it was present in cultural liquids at development of bacteria on the nutrient mediums which are not containing cellulose (the glucose-peptone nutrient medium, the nutrient broth).

At development on various nutrient mediums cells *B.subtilis 8130* allocated metabolites, stimulating in modeling systems development of tests- cultures: lactobacillus, lactic streptococci (fig. 2, table 2).

Table 1. Endoglucanase activity of culture *B.subtilis 8130* on various nutrient mediums

1	Kind of test	Number of cell / ml N x 10^9	Activity, µmol/min/ml	Specific activity, A/cell x 10^{-9}
1	The sterile glucose-peptone nutrient medium	0	0	0
2	The sterile nutrient medium with corn flour and barmy autolysat	0	0	0
3	The sterile nutrient medium with corn flour, soya flour, sunflower presscake	0	0	0
4	Culture on the nutrient broth	2	3.8	1.9
5	Culture on the glucose-peptone nutrient medium	2.5	12.5	5.0
6	Culture on the nutrient medium with corn flour and barmy autolysat	3	7.4	2.5
7	Culture on the nutrient medium with corn flour, soya flour, sunflower presscake	1.5	7.4	4.9
8	Culture on the nutrient medium with beet bagasse and corn gluten	2	13.7	6.85

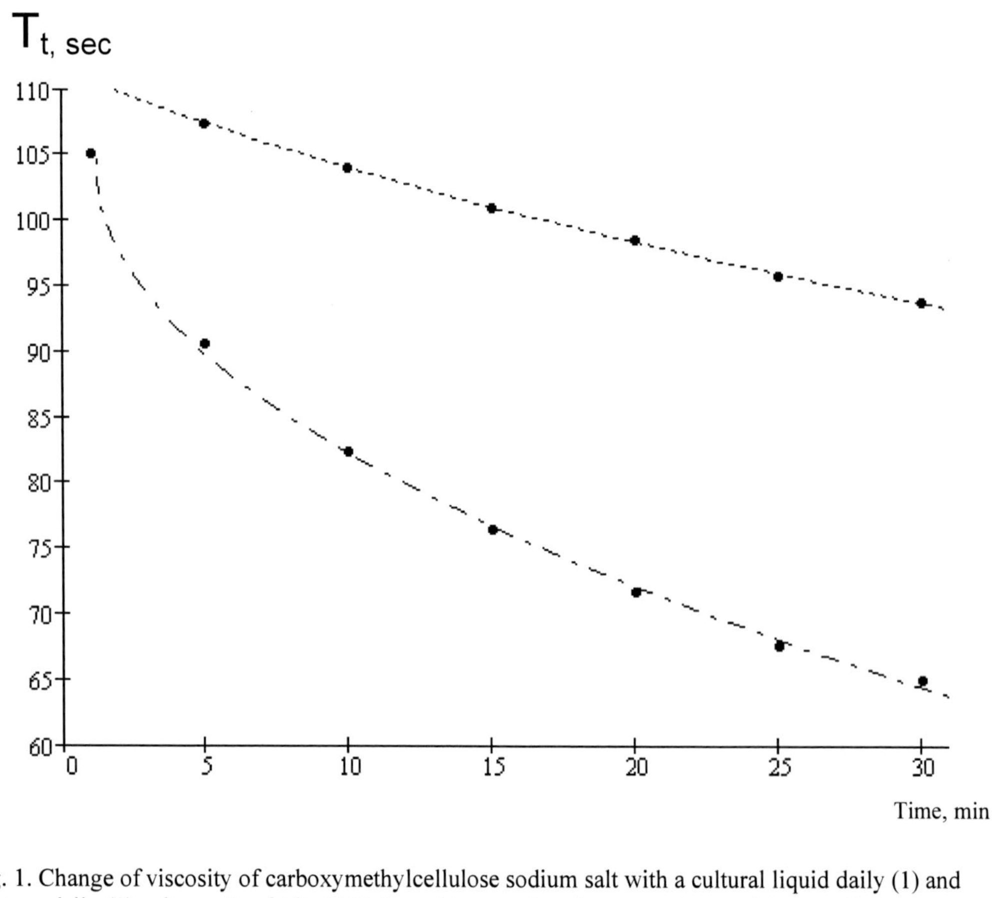

Fig. 1. Change of viscosity of carboxymethylcellulose sodium salt with a cultural liquid daily (1) and the two-daily (2) cultures *B.subtilis 8130*, brought up on the glucose-peptone nutrient medium

Table 2. Parameters of an index of stimulation of lactobacillus by daily cultural liquid *B.subtilis 8130* which have been brought up on various nutrient mediums

Nutrient medium	Index of stimulation
The glucose-peptone nutrient medium	2.5
The nutrient medium with corn flour, barmy autolysat and sunflower presscake	2.0
The nutrient medium with corn flour, soya flour and sunflower presscake	2.5
The nutrient medium with beet bagasse	1.5
The nutrient broth	2.5

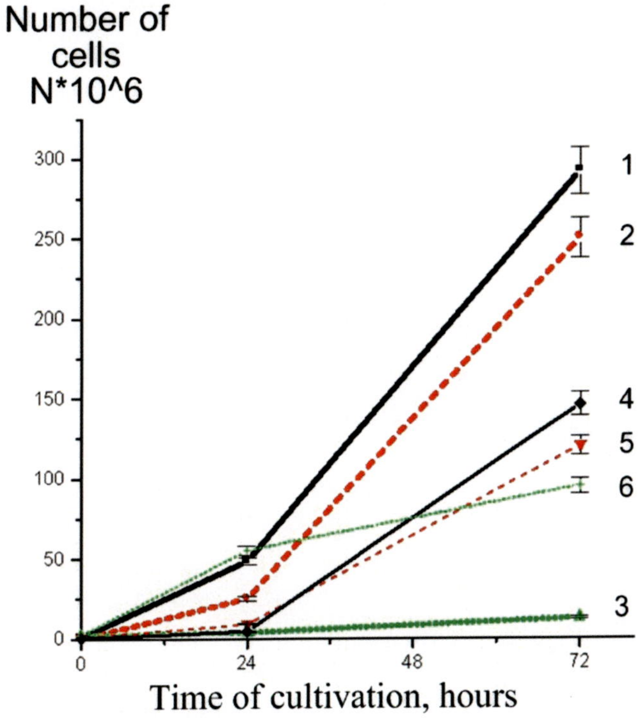

Fig. 2. Change of number of cells at joint development of tests - cultures with *B.subtilis 8130* : 1 - *Lactobacillus acidophilus Suav. B-4625*, 2 -*Streptococcus lactis*, 3 - *B.subtilis 8130*. Controls: (separate cultivation of bacteria): 4 - *L.acidophilus Suav. B-4625*, 5 - *Str.lactis*, 6 - *B.subtilis 8130*

Allocation by cells biologically active metabolites increased in parallel with increase of number of cells (fig. 3). However, at development on a nutrient medium with cellulose (sunflower presscake) the stimulating factor by 30 hours of cultivating has reached the maximal value, as against endogluconase activity which has essentially risen by 65 hours.

The stimulating factor and endoglucanase activity are consistently allocated with cells, providing ability to live of bacilli in microbiota, and, apparently, are differently connected to sources of nitrogen and carbon. So, for example, on the nutrient medium with beet bagasse on which it is high endoglucanase activity, stimulation is expressed poorly. On nutrient broth, on the contrary, at a low parameter endoglucanase activity stimulation is obviously expressed. It is possible to assume, that the stimulating effect depends on a source of nitrogen, and its allocation is determined by enzymatic hydrolysis of peptides, that causes stimulation lactobacillus in a trophic circuit intestinal microbiota. Endoglucanase, on the contrary, is functionally connected to accessible sources of carbon and with decrease in the maintenance of easily utilized sugars resulting in increase of synthesis of enzyme, intended for splitting cellulose.

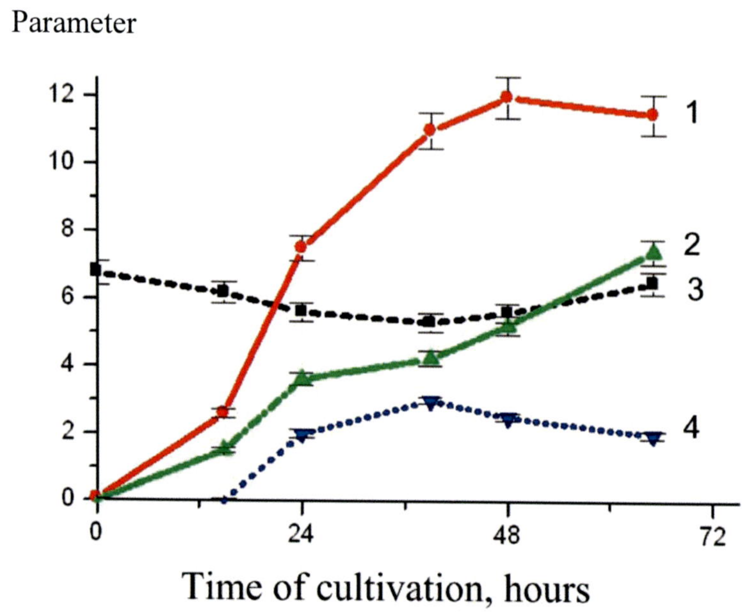

Fig. 3. Parameters of development of culture *B.subtilis 8130* on the nutrient medium with corn flour, soya flour and sunflower presscake. 1-number of cell / ml x 10^8, 2-endoglucanase activity, a unit, 3-**pH**, 4-index of stimulation of lactobacillus

For revealing biological efficiency metabolites of *B.subtilis 8130* in forages of young dwarfish rabbits 24 and 48 hour cultures *B.subtilis 8130* on the corn-soya nutrient media with sunflower presscake have been entered on 0.1 % from weight of cultural liquids. The average additional weight of skilled rabbits on a diet with daily cultural liquid in which the index of stimulation made 2.5, and endoglucanase activity - 3.6 units, has been comparable with an additional weight of control animals, and has made, accordingly, 21.34 g and 22.95 g (the difference is statistically doubtful, $P > 0.05$). On a diet with two-daily cultural liquid (endoglucanase activity - 7.4 units, an index of stimulation - 2.0) the additional weight of skilled animals on 11 % was higher, than control, and has made 25.48 g ($P<0,1$). Hence, the combination is necessary for reception of positive influence on an organism of animal stimulating metabolites with endoglucanase activity.

The stimulation role of strain in digestion of animals is shown in experiences on mixture *B.subtilis 8130* with a dry beer pellet (13-16% cellulose) and to introduction of 6 % such enriched by bacilli pellets in a diet of calf (table 3). At animals has increased digestion organic substances of forages, the quantity of the general fiber of whey of blood and postponed in a body of animals of nitrogen has increased, and as a whole 12 % of an additional gain of alive weight are received at 10 % of economy feed cost.

Table 3. The basic results of experimental feeding calf mixed fodders with a dry beer pellet and probiotic *B.subtilis 8130*

Parameters	Groups		
	Control	1 skilled	2 skilled
Alive weight, kg: in the beginning of experience	53.5±2.78	53.7±3.10	53.6±1.43
At the end of experience	124.5±8.11	125.1±3.33	133.1±4.37
Gain of alive weight, kg	71.0±7.7	72.4±3.5	79.5±4.3
Daily average gain of alive weight, g	772±84	787±38	864±47
Too in % to the control	100,0	101.9	111.9
Expenses for 1 kg of a gain:			
Exchange energy, MJoule	33.8	34.0	31.8
Mixed fodders, kg	2.24	2.20	2.01
Too in % to the control	100,0	98.2	89.7
Digestion, %:			
Dry substance	73.3±2.46	73.7±1.48	75.3±0.65
Organic substance	74.6±2.36	75.1±1.42	76.8±2.14
Balance of nitrogen, g/day	23.2±0.91	25.8±1.43	26.8±2.14
The general fiber of whey, g/l	63.4±1.4	65.1±2.1	67.4±1.6

Distinctions are statistically authentic in comparison with the control over value P <0,1.
Control group - mixed fodder without a beer pellet
1 skilled group - mixed fodder about 6 % of a beer pellet
2 skilled group - mixed fodder about 6 % of a beer pellet with *B.subtilis 8130*

Stimulating splitting cellulose activity of strain *B.subtilis 8130* has allowed to develop a way of activation of the strain in the way feed to a beer pellet with bacilli to an earthworm. After a month of the maintenance of earthworm on a dry beer pellet with *B.subtilis 8130,* a strain of *Bacillus* has been allocated from intestines of a worm. This strain on morphological, cultural and to physiology-biochemical parameters similar with collection strain *B.subtilis 8130*. The DNA fngerprint has shown identity strains. At strain, allocated of a worm, endoglucanase activity on environments with cellulose was higher 15 %, and it differed by increased antibiotic action.

The received data testify to participation of the representative of kind *B.subtilis* directly during digestion as the natural catalyst.

REFERENCES

[1] I.A. Poberij, N.V. Litusov, A.V. Stepanov, O.P. Misnikov, A.N. Zabokritskij, P.G. Vasiliev: Probiotics, prebiotics, sinbiotics and functional food stuffs. *A modern condition and prospects.* The collection of materials of the international conference. Moscow, Rossija. On June, 2-4. 2004, 72 (in Russian).

[2] N.A. Mihajlova, M.A. Godkov, A.G. Gataulin, V.J. Zinkin, A.I. Vetoshkin, A.V. Haritonova, A.A. Gaidarov: Probiotics, prebiotics, sinbiotics and functional food stuffs. *A modern condition and prospects*. The collection of materials of the international conference. Moscow, Rossija. On June, 2-4. 2004, 204 (in Russian).

[3] X. Bergner, X.-Ç. Ketz: Scientific bases of a feed of agricultural animals. Kolos, Moscow, 1973. 567p. (in Russian).

[4] I.N. Gritsenko: Microflora fine mammal Western Siberia. *Science*, Novosibirsk. 1971. 305 p. (in Russian).

[5] E.A. Orlova: In *Termiti*. The Moscow State University, Moscow, 1972. 157p. (in Russian).

[6] N.A. Ushakova, L.P. Belov, A.A. Varshavskij, A.A. Kozlova, T.V. Kolganova, E.S. Bulygina, T.P. Turova: *Microbiology*, 72 (3), 400 (2003).

[7] N.A. Ushakova, D.S. Pavlov, B.A. Chernuha, M.P. Kirillov, A.J. Jahin, O.V. Sokovyh, A.A. Kozlova, A.V. Nifatov, A.R. Abdrafikov: Probiotics, prebiotics, sinbiotics and functional food stuffs. A modern condition and prospects. *The collection of materials of the international conference.* Moscow, Rossija. On June, 2-4. 2004, 162 (in Russian).

[8] Bergey's Manual of Systematic Bacteriology // Williams and Wilkins. 2, 1984. 1105.

[9] A.P. Sinitsyn, A.V. Gusakov, V.M. Chernoglazov: *Bioconversion of lignine-cellulose materials*. The Moscow State University, Moscow. 1995. 146 (in Russian).

[10] N.S.Egorov (Ed.): *A manual to practical employment on microbiology.* The Moscow State University. Moscow 1983. 215 (in Russian).

In: Biotechnology, Agriculture and the Food Industry
Editor: G. E. Zaikov, pp. 119-125
ISBN 1-60021-040-6

Chapter 20

A GREAT NATIONAL PROPERTY OF RUSSIA FERTILIZERS OF XXI CENTURY – "FLORA-C" AND "FITOP-FLORA-C"

N. M. Boyarskiy*
CLR "BIO-BAN"

From time immemorial mankind shows a great interest in the problem of raising soil fertility and its restoration after natural cataclysms and irrational activity of a man.

In Egypt and in the valley of Nile at the very outset of civilization people have already used humic acids (ooze contained humic acids with concentration of 2 g/l. And already at this time people used a fertile layer of the valley of the river for improving the nutritious properties of soil. Later on, scientists proved that in formation of soil fertility humic acids were of great importance, preparations with the content of humic acids 2g/l have been developed. For the first time in Russia humic acids have been received in 1936. Till 1995 concentration of humic acids remained to be low – 2 g/l.

Nowadays, there are preparations with higher concentration of humic acids. At present the leader in quality of peat-humic fertilizers is CLR "BIO-BAH" - the laureate of 100 Best Goods of Russia in 2000-2004, producing ecologically pure dry peat-humuc fertilizers "Flora-C" and "Fitop-Flora-C'.

DPHF "Flora-C" is a highly concentrated mixture of biologically active substances, isolated from ecologically pure natural raw material, balanced in the content of micro- and macro-elements with high content of humic acids (not less than 12 g/l). The property of "Fitop-Flora-C" is analogous, but it contains stock of registered monobacterium, which is effectively fighting with rotten micro-flora, working out more than 70 peptide antibiotics-bacitracins, and also increases the quantity of bacteria, which are of great importance in ecological state of humus.

Preparations have the stable composition, these are specific biological preparations, because the stock, that we use, is a natural stock, which doesn't die, as the medium, where it

* Our address: Pochtovyi 12, Biisk, 659300, Altai territory, Russia. Tel./fax. (3854)-32-93-21, E-mail: info@bio-ban.com, www.bio-ban.com

lives, is ecologically pure, not aggressive. It does not open, because pathogenic rotten micro-flora is absent in preparations.

DPHF "Flora-C" is a fundamental preparation. "Fitop-Flora-c' – a preparation, which is a base not only for plant-growing, but for other branches of agriculture. Now many firms try to develop fertilizers with a great quantity of bacteria, which are contrary to each other. You should be careful, when using such preparations. When pathogenic micro-flora is killed, bacteria begin to kill each other, because there is the law of nature, where the strongest survive. It can lead to bacteria mutation and as a result of it, to biological war in nature. In contrast to these preparations, our preparation contains monobacterium, (one stock), providing for preparation "Fitop-Flora-C" with a protective action of plants and soil, raising resistance to different diseases: phytophthora rot, gray rot, mildew, downy mildew, phomosis, blackleg, rhizoctoniose and so on.

What Differs Us from other Producers of Organic Fertilizers?

DPHF "Flora-C" and "Fitop-Flora-C" have no analogs in Russia and abroad, they are protected by patent of the Russian Federation. The preparations are included in State register of the RF, admitted for application in Russia, approved for application by the ministry of health services and International System of Quality (USO 9000). "Flora-C" and "Fitop-Flora-C" satisfy the demands for application on all the continents, in all natural-climatic zones. Proceeding from sanitary conclusion, our preparations are inferior to the demands, made of food products, in ecological purity.

1) DPHF "Flora-C" and "Fitop-Flora-C" are safe preparations for a man and environment.
2) Our preparations have a high concentration of humic acids – 12 g/l. As is generally known, the base of fertility is humus, the base of humus - humic acids. Humic acids, being a part of "Flora-C" and "Fitop-Flora-C", refer to the 6-th generation of humic acids, which are characterized by high concentration, biological activity, purity and ability to be assimilated by animate and conditionally animate nature. Humic acids are natural catalysts, and there are up to 60% of insoluble substances useful for plants, in soil, which are accumulated in over clay layer of soil. Humic acids as natural catalysts, make them biologically active.
3) By making of the given preparations we, for the first time, achieved the united biochemical and microbiological process:
    (a) preparation of soil and seeds – if the quantity of humic acids in soil is reduced, the yield will be low, because rotten micro-flora will be propagated. Preparations increase the number of Azotobactera, Fosfobactera and Bacillus subtilis, playing an important role in ecological state of the ground; if the seeds are affected by pathogenic bacteria, after soaking them in solution with PHF, the seeds won't germinate, as humus acids will prevent from it.
    (b) root and foliar fertilizing.

(c) treatment of plants after application of chemical protection means (humic acids are natural absorbent, and so they neutralize the excess of chemical protection means of the plants in soil). And as a continuation, treatment of storehouses and root-crops before continuous storage.

4) One more distinctive character of our preparations – the absence of derivatives of the red Californian worm, because the last one leads to soil disturbance. Soil nutritious properties are lowered, excessive water-permeability appear and also we don't use chicken and fowl manure, as the manure increases the amount of rotten micro-flora in soil, which affects the plants, and at continuous storage of manure in a moist state or at freezing, nitrogen is being lost.
5) It is very important to note, that We don't work with potassium humate. In industrial production for obtaining potassium humate brown coil is used, which is a natural absorbent of heavy metals. Under the influence of factor of time heavy metals interact with humic acids. Chemical absorption of potassium humate by a soil leads to such phenomenon, that humic acids transfer heavy metals in active state This process is ensured by transport and protective function of humic acids. Carried out investigations showed a low concentration of humic acids in potassium humate. Proceeding from the reasons, given above, in some civilized countries potassium is forbidden for application. The base of our preparations – sodium humate, one of the components of which is natural peat. Sodium humate has high concentration of humic acids, which transfer heavy metals from biologically active form into inert one, bind toxic and radioactive elements into not mobile and hardly dissociating connections and also connections, negatively influencing in ecological safety of nature, including they can incorporate some pesticides, hydrocarbon, Phenolums. The humate of natrium demands the small consumption.
6) The laboratory expertise spent in different countries, completely have confirmed all characteristics of our preparations. DPHF "Flora-C" and "Fitop-Flora-C" have shown the universality at application in extreme situations (at reestablishment of fertile layer of soil after fires, flooding, earth flows). In regions of ecological calamities preparations reduce sanitary-epidemiological tension. With this purpose it is possible to apply DPHF for decontamination of the settlement, dumps, waste from animal industries, realization of preventive actions on cattle-breeding complexes.

Application of the given preparations increases seed germination by 3-5%, energy of seed sprouting by 4.5-18.5%, promotes increase of a root system in 1.7-2.5 times , shortens ripening time by 10-15 days, prevents from accumulation of biologically injurious substances in fruits, increases the yield by 20-40%, improves stability of plants to a drought, frosts, excess of chemical means of plants protection in soil, promotes increase of humic acids in soil (where they don't allow the process of rotting, actuating process of decomposition), transfers heavy metals from the biologically active forms into inert one, reduces soil acidulation, rich in iron (red ferrallitic), promotes soil lamination (on solonetzs, salines), improves buffer properties of soil, aeration on waterlogged soils and in regions of a permafrost, accelerates introduction and breeding processes of species, improves the structure of soil, suppressing the development of autochthonous micro-flora, destroying humus.

Except general instructions in application of the given preparations, there are separate methods of agrotechnics:

The method of discontinuance of sands movement by creation of humus horizon (the humic acids transfer inter-oozy connections, located in sans, from the biologically passive form in active. The process of formation of fertile layer of soil takes 12-24 months, and in conditions of non-irrigated agriculture – 24-36 months.

As a consequent, establishment of a peat lawn and landscape design on sand soil (time of establishment of a steady peat lawn in 12-24 months from the moment of formation of humic horizon).

Technology of recultivation of oil-contaminated grounds, that is very important at a biological stage. The application of the given preparations allows in short times (less than year) to restore a fertile layer after an oil spill, to increase germinating capacity of seeds in oil-contaminated soils, shorten terms of sprouting and to increase air-dry weight of plants more than in 10 times.

The technique on reestablishment of fertile layer of soil after flooding, fires and earth flows during 8-12 months is designed at the expense of acceleration of biochemical and microbiological processes in soil. Average consumption of peat-humic fertilizers is 2-4 kg for 1 ha. At repeated application of fertilizers in the same volumes terms of reestablishment are shortened up to 3-6 months. The application of mineral fertilizers for soil reestablishment, at first, is non-deficient (600kg/1 ha), secondly, will not bring a desirable result (they do not influence in soil fertility).

The new technique of cultivation of sugar-beat from seeds is designed. Thus, the area under crop in 3 times is reduced, the yield increases (at single treatment with DPHF up to 280 c/ha), as well as sugariness, stability to root and clamp rot in stationary storehouses, the intensity of development of root-eater is reduced by 44.6%, the safety of a crop is provided by 85-90%.

The yield increase at field tests has compounded: summer wheat – 15-25%, carrots – 25%, fodder beet – 25%, corn – 20-40%, early cabbage – 29-35%, garden beet –35%, cotton plant – 8-13%, tomatoes in closed ground – up to 40%, radish – 30%, soy-bean –20-27%.

The tests have shown, that applying DPHF “Flora-C” and “Fitop-Flora-C”,at planting, terms of taken roots are shortened by 3-5 days. For example, applying DPHF “Flora-C” and “Fitop-Flora-C”, it is possible to shorten terms of taken roots, when planting seedlings of cucumbers.

We have a technology of obtaining of humus acids of high scale of purity. All listed above and other techniques and the technologies have scientific-practical endorsement of many Universities and other organizations, specializing on these problems.

What Is Necessary for Obtaining a Qualitative and Rich Crop?

1. Proper Preparation of Soil and Seeds Very Important

The application of our fertilizer makes soil fertility, improves soil structure. Good results on application of DPHF”Flora-C” are obtained in soils even with the lower content of nutrient materials: sandy, sandy-loam, clay, soddy podzolic, wood etc.

The obtaining of high crops of different agricultural crops is possible only at strict observance of definite balance of organic and inorganic substances in soil.

Abusing of mineral fertilizers and the full absence of attention to organic part of soil results in eco-catastrophe, decrease of the yield, deterioration of ecological purity of agricultural production, it's flavouring qualities. For solving of the given problem the application of natural organic fertilizer DPHF "Flora-C" and "Fitop-Flora-C" is offered, which one, on the one hand, allows to receive ecologically pure production, and, on the other hand, makes soil healthy, to improve it's structure, restoring thereby natural fertility of agricultural lands. Besides, DPHF "Flora-C" and "Fitop-Flora-C" will actuate vital activity of different groups of microorganisms, favourably influencing on the growth and development of plants.

The best result is achieved at seeds soaking in DPHDF "Flora-C" or "Fitop-Flora-C" for 24-72 hours. DPHF "Flora-C" and "Fitop-Flora-C" are fertilizers of prolonged action and can be applied for all agricultural crops. Positive influence of DPHF "Flora-C" on plants is expressed in acceleration of seeds germination. What highly productive the seeds were, but if the soil is poor and, therefore, is affected by pathogenic bacteria, it is useless to wait for a rich crop. We have never seen a healthy child, born from a sick mother. The tests, carried out in a zone of a risky agriculture, have shown, that the application of DPHF "Flora-C" and "Fitop-Flora-C" promotes a survival of plantlets, strengthening of a root system in 2.5 times, stimulation of respiratory processes, photosynthesis, acceleration of development of plants, as results in increase of their productivity.

2. Root and Foliar Fertilizing

One packet of "Flora-C" 30 g is used for foliar fertilizing for 500-600 sq. m., root fertilizing – 80-110 sq. m. One packet of "Fitop-Flora-C" (10g) is sufficient for foliar fertilizing for 160-260 sq. m, root –30-60 sq. m, and for treatment of green houses, storehouses and root-crops for 150 sq. m.

At commercial application consumption of fertilizers is 1 kg for 2-3 ha – for soil preparation and root fertilizing, 1 kg for 7-10 ha – for foliar fertilizing, treatment after application of chemical means of protection of plants.

At personal plot, three packages of "Flora-C" and two packages of "Fitop-Flora-C" is necessary for $100m^3$ from spring to autumn, on the average.

For treatment of a greenhouse with an area of 30 sq. m. one package of "Flora-C" is required – for soil treatment and five-six packages of "Fitop-Flora-C" – for soil preparation, foliar fertilizing and treatment of a greenhouse.

The detailed description of a preparation and application of solutions of DPHF "Flora-C" and DPHF "Fitop-Flora-C" is given in instructions. A liquid solution of preparations is applied in soil and on plants in the following way:

1) in a garden – with the manual sprayer.
2) on the large areas – with the help of aggregates ОПВ-120, ПОМ-630, "Kirkitoks" and similar to them.

3. Treatment of Plants after Application of Chemical Protection Means of Plants

It is necessary to mark, that as a result of single treatment of soil with DPHF "Flora-C", the content of radionuclides in agricultural production decreases in 2-3 times at cultivation on soils, contaminated by radioactive isotopes of cesium – 137 and potassium – 40, since the humic acids are a natural absorbent, therefore, they neutralize excess of chemical protection means of plants. DPHF "Flora-C" allows to lower the content of nitrates in agricultural products in 10-30 times and the content of radionuclides in 2-3 times, to improve soil structure, to increase gas-and water permeability, to observe the necessary proportion of a structure of mineral fertilizers and nutrient biological substances.

Now in agriculture for protection of plants against pests and diseases chemical agents are the main means. For many years of application of chemical weed-killers and pest-killers the essential damage is marked to nature, natural balance between a useful pathogenic micro-flora is disturbed: killing parasitic insects, killed hundreds and thousands useful. For pests, carriers of diseases and pathogenic microorganisms, the immunity to many poisons was worked out. Working out of new poisons results in occurrence of more and more steady microorganisms, and hereinafter in reply to superpoisons there are supermicrobes. As the director of the research-and-production corporation "Research center" A.I. Lelyak says: "Microbes – surprisingly perfect and self-teaching machines, which are capable to "remember" gears, worked out by them for protection against destructive action of poisons". Therefore, we go by a way, built by the animate nature for millions of years of evolution of a microcosmos. And in nature all is so, that the pathogenic micro-flora loses in flight with useful bacteria more often. Differently, on the Earth there would be no such rich diversity of every living thing, including a man. Our way is to find useful bacteria in nature, "to educate" them, to work out preparative forms of new stocks and to put them on a service to a man, agriculture and protection of the nature

We achieved great results in greenhouse complexes, the main problem of which is pathogenic, rotten micro-flora. Actively developing in soil, during condensation it raises on constructions of greenhouses. When the greenhouse is moistened, bacteria is directed downwards and affects plants and soil.

The application of PHF "Flora-C and "Fitop-Flora-C" allows:

- to lower sharply the quantity of pathogenic (parasitic) microorganisms in soil,
- to restore a biochemical and microbiological structure of soil,
- to lower sharply the quantity of toxins,
- to restore biological balance in soil,
- to restore an ecological state of a humus,
- to prevent from affection of the plants with rotten micro-flora.

DPHF "Flora-C" and "Fitop-Flora-C" are applied for cultivation of fruit-berry, vegetable crops, cereals, grass, flowers and other plants in the open ground and greenhouses. The given fertilizers are available for cultivation of indoor plants, establishment of lawns, flowerbeds and other activities in landscape design.

Taking into account a small volume of soil in a pot or tub of the indoor plant, which becomes poor in time, it is necessary to water soil with a nutrient solution.

Lawn grasses, flowerbeds also require additional nutrient materials, protection against unfavourable weather phenomena and affection by different diseases.

The application of DPHF "Flora-C" and "Fitop-Flora-C" allows:

- to increase nutrient, buffer properties of soil,
- to improve structure and gas changing of soil,
- to shorten terms of taken roots of plants at their replanting, propagation at the expense of development of optimum biological and microbiological medium,
- to increase the duration of the active period of indoor blossoming, ornamental-leafy, ampel, climbing plants, succulents, to stretch the period of vegetation of lawn grasses, flowerbeds,
- to increase the energy of sprouting and germinating capacity of seeds,
- to make strong a leaf on plants,
- to lower toxicity of soil, to raise immunity of indoor plants, lawn grasses, flowerbeds to different diseases.

For high quality, ecological purity, efficiency and economy DPHF "Flora-C" and "Fitop-Flora-C" have become the Laureates and owners of sign of Programmer "Russian Quality", three times Laureates of the Programmer "100 Best Goods of Russia", owners of quality Sigh, golden medal of the competition "Ecologically Pure and Safe Products", preparations are awarded with 21 Golden medals of different exhibitions-competitions, including international ones.

Before application of our preparations, learn directions attentively and follow them, only in this case, you'll receive an abundant and healthy crop.

We advise you to pay attention to a label. There, except the name of a preparation and corporations-manufacturers, we point numbers of state registration (N° 26-8051-0175-1), (26-9023(9024-9027)-0175-1) and registration numbers of a tare label by all means.

In: Biotechnology, Agriculture and the Food Industry
Editor: G. E. Zaikov, pp. 127-131
ISBN 1-60021-040-6

Chapter 21

Role of Method of Chemical Mutagenesis in Creation of Biological Diversity and the Souses of Rare and New Characters in Winter Wheat

Natalya S. Eiges, Larissa I. Weisfeld and Georgij A. Volchenko
Emanuel Institute of Biochemical Physics, Russian Academy of Sciences,
Kosygina str. 4, 119997, Moscow, Russia

At the present time in Russia, the biological diversity of different crops is reduced, especially of wheat, as the most widely distributed crop. This loss of diversity of crops is the result of reducing diversity of genotypes taking a role in creating new varieties as a result of wide expansion of few varieties, named varieties-hegemonies, even if they are vary valuable and deserve wide expansion. In works of creating of this few varieties take role genetic monotonous initial material by intraspecific (intervarietal) crossings. As a result, intervarietal hybridization decreases and its possibilities are exhausted in many respects.

Similarity of genotypes of different varieties and wide expansion of varieties-hegemonies are dangerous from the point of view of the quick expansion of new virulent races of phytopathogens and of development of epiphytoties even on formerly resistance varieties. For example in the sixties of XX. Century epiphytoty of brown rust of wheat destroyed valuable resistant varieties Aurora and Caucasus, created by P. P. Lukyanenko. Genomes of these varieties contained in many respects similar genetic material of variety Bezostaya 1, created by Lukyanenko also. Varieties, expended in the South of Russia, contained predominantly genetic material of variety Bezostaya 1, varieties, expended in the middle of Russia, — of variety Mironovskaya 808.

Epiphytoties arise especially frequently among unresistant varieties. But resistant varieties with monogenic resistance expose to epiphytoties also, in the event these varieties are expended widely, especially towards to pathogens with quickly rise of new more virulent races, for example to brown rust of wheat. Also epiphytoties affect varieties with monogenic resistance to pathogens with slow development of races, for example to stinking smut. Thus in the thirties wide expanded in Europe variety of wheat Omar left the stage, affected by epiphitoty of stinking smut.

Also it is dangerous wide expansion of few in number of varieties with similar genotypes at unfavourable conditions even for adaptive varieties in a given region. It is necessary for the saturation of each region by diverse genotypes for the avoidance of epiphytoties and for the guarantee of stable yields in different years including extraordinary conditions. It may be achieved by means of mosaic arrangement of varieties with diverse genotypes.

But how possible to achieve this diversity of genotypes in case present varieties are uniform and it is absent sources of several necessary for selection valuable characters? At the beginning of the twentieth century the now losing diversity was provided in Volga region by aboriginal varieties of spring wheat. In the experimental station near Saratov (now Research Institute of Agriculture of South-East) hybridization of wheat begins between aboriginal south-east varieties with high bread-making characteristic.

On the basis of local aboriginal varieties selection was development so energetic, that it was created by only one selectionists A. P. Shehurdin 30 genotypic diverse varieties for providing other regions not only Saratov region. These varieties make significant contribution in high yields and high quality of wheat. This is one of causes, why at the beginning of former Central part of Russia occupied one of first places on the production and export of kernel of wheat.

For development of selection it is essentially to revive genotypic diversity of wheat on the new level — with higher crop capacity.

For increasing of genotypic diversity of spring and winter wheat and receiving of valuable new and rare characters exists of method of distant hybridization of cultivated varieties of wheat with wild species, for example, with *Triticum monococcum, Tr. dicoccum, Tr. timopheevi*, and with other genus — rye, *Aegilops, Agropyron*. In guaranteeing of sources for distant hybridization plays a role N. I. Vavilov all Russian institute of plant industry.

By means of distant hybridization such characters as resistance to brown rust were passed to wheat from different species of *Aegilops*, and resistance to powdery mildew from *Agropyrum glaucum* (*Elytrigia trichophora*). Diverse genetic material including new and rare characters brought in genome of wheat by such crossings. In this way the given sources of wheat characters arise by utilization of intermediate forms by means further hybridization. Method of distant hybridization was given many positive results. Varieties of winter and spring wheat of food-producing and forage purpose created by N. V. Cicin, who for the first time used crossing of wheat with different species of *Agropyron*.

However given method very laborious and long. In generations of these hybrids it remains sterility, absences constancy, especially when the species or genus differ by number chromosomes. Besides undesirable characters constantly go with desirable characters, passed from other more frequently wild species or genus, closely associated with need characters.

Liberation from undesirable characters require some backcrossings that one not always effective although they allow for the application of cellular or genetic engineering.

Genetic diversity can be received be intervarietal (intraspecific) crossing in the event that the parents represent differ geographically distant varieties, which have essential differences in genetic constitution.

Agriculture of our country is situated in a difficult economic position. As a result it was lack or absence of machinery, fertilizers, herbicides, and pesticides. Rotation of crops, incorrect using of agricultural chemicals and mineral fertilizers, limited quantity or absence of organic fertilizers get broken. These brought to disturbance of microbiological processes in soil and worsening of its fertility and structure.

The correlation of two most important factors takes the role by plant-growing: a factor influencing the on phenotype of the plant — technology of cultivating, and second factor genotypic, responsible for totality needing valuable for selection hereditary characters — it is variety.

In variously time correlation of role of these two factors changed. If quantity of varieties were not enough, the factor technology overweighed. This way was at beginning XX. Century. In that case quality of organic fertilizers, cultivating on fallow, enough high agricultural background, rotation of crops, not disturbed microbiological processes, structure of soil, absence of herbicides and pesticides played the topside role. Also it was impossible enough number of agriculture people and absence modern techniques, absence of large agriculture. At the same time were more or less stable climatic conditions for zones growing of wheat, against a background of which living many years worked out technology used.

In the course of time the correlation of role of technology of cultivating and of role of genotype of variety changes. Little by little the climate changes. For example climate in the Climate in the Central region of Russia became less favourable for cultivating winter wheat during last ten years. Now number of people occupied in agriculture is decreased. The role of techniques, pesticides, and herbicides is increased, the volumes of organic fertilizers came down, and mineral fertilizers appear, which displace organic fertilizers more and more. At present time the volumes of the one and the other fertilizers is decreased. Rotation of crops, microbiological processes and structure of soil is changed.

By change of climate and technology of cultivating of wheat (principally to unfavourable aspect) role of variety increases more and more and still greater requirements claims to variety. Thus role of selection increases. Difficulty of modern selection of wheat and its development is connected not only with economic difficulties of science, but also with other causes.

One of them consists of decreasing intraspecific (intervarietal) hybridization, which in former times was highly effective. Basic cause of that consists in decreasing genotypic and phenotypic diversity. Dynamics of the decreasing increases. Decreasing of biodiversity created by wide expansion of not numerous genotypes of varieties-hegemonies. At the same time similar genotypes is repeated in new initial material for selection and in varieties. Assortment of initial material of new varieties is impoverished and has limit.

Monotony of assortment is increased because the species of winter wheat *Triticum aestivum* L. have not all the characters needed for selectionists. For example, it has not the characters of resistance to powdery mildew and stinking smut. Not numerous lines and varieties resistant to the phytopathogens acquired this resistance from other species or genus by distant hybridization. Distant hybridization is vary long and labour-intensive and not always effective. Much simpler to get the absentee at wheat new characters of resistance by method of induced mutations, in the first place is method of chemical mutagenesis as more effective.

Genotypic and phenotypic monotony because of impoverished initial material for selection increases also the fact that they are found rarely at present species of wheat and to receive their vary difficult at re-created varieties be traditional methods of selection and to keep their at descendants is else more difficult. To rare characters belong such valuable characters as high content of protein in grain, high bread-making quality. Usually the characters depend on environment more then on genotype. It is vary difficult to get the characters genetic conditional as by intervarieties crossings so by method distance hybridization.

Biodiversity also limited by correlations between the characters, that as a rule are not overcome by usually methods of selection.

Danger for decreasing biodiversity is transgenic plants if they will introduce. We think that now possibilities of method of chemical mutagenesis are enough great for increasing biodiversity and decision in a number of problems above mentioned. By working with induced mutagenesis it is important in order that mutagene would be effective, and the variety would be mutable. Most important is rightly selection of combine of mutagene, of diapason of its doses and initial variety of wheat. By optimal combine this three components it is not obligatory each from them will be individually in such a way effective. For example, super mutagenes nitrosomethyl urea (NMU), nitrosoethyl urea (NEU) discovered by I. A. Rapoport after discovered by him ethylene imine (EI), and are counting more effective in comparison EI. Opinion exists that it follows as initial variety by induced mutagenesis to use the variety that is best for given area or districted (or included in State register). However it is prematurely to throw out on genetic dump the varieties that more not cultivated. Some of these varieties may find oneself the most perspective by application them as initial material by combination with that or either chemical mutagene.

For example, in our case from four tested mutagenes using on the tree varieties it was the most perspective combination mutagene EI + the diapason of its doses 0,01—0,04% + initial variety — hybrid wheat-agropyrum PPG 186. This variety is not cultivated now in Central region starting with sixtieth past century. However the EI tested on the varieties Mironovskaya 808 (that at present time substitutes variety PPG 186), PPG 599, PPG 1 and others did not give such strong effect as tested on variety PPG 186. This is related as width of spectrum of mutative changes both to the frequency of mutants with economically valuable characters. After investigations we also refused from combinations of NMU, NEU, diethyl sulphate, dimethyl sulphate with other varieties besides combination EI with variety PPG 186.

Method of chemical mutagenesis unconditionally showed oneself as one of the most effective one for receipting of high frequency of mutations, the theirs wide diversity, and high going out of economic valuable mutations. But potential possibility of this method did not yet exhaust. In particular great investigations are in prospect for different combinations of chemical mutagene, its doses, different crops, their varieties, and choice of the most effective combinations. It is not except that it would found yet more favourable combinations in comparison with our data. The most successful in given investigation is combination of ethylene imine in doses 0,01—0,04 volume % with the variety PPG 186 by treatment of air-dried seeds during 24 hours.

The frequency of mutations is high and by different methods of estimation represent:

- 10—12% mutant plants in respect of total number of investigated plants in second generation (M_2);
- more then 50% families with mutants in respect of total number of investigated families in M_2;
- more then 100% mutations in respect of total number of investigated families in M_2.

Similar frequencies of mutations was observed also by other combinations of mutagenes, theirs doses, and other initial varieties. But such high genotypic and phenotypic diversity of mutations and mutants was received for the first time only in given work.

Everything diversity is composed by 50 types of mutants. At the same time the number of types of mutants was not higher then 9—13 in other works by employment of more effective mutagenes and other varieties. By optimal combination on mutagene EI and its optimal doses with initial variety PPG 186 every selected mutant had a valuable character or valuable characters. Thirty percentages had only one valuable character or complex of them without accompaniment of negative characters. These 30% of mutants are not needed in selection elaborate. They are valuable initial material, which may be passed for State tests of selection achievements immediately after reproduction or after negligible number of selections. On the basis of mutants received by treatment of EI it was created the large collection of similar mutants. Part of the mutants of this collection, which were received by treatment of the initial variety PPG 186 by optimal doses of EI, have no breaches of chromosomes but have gene mutations controlling valuable characters heritable in generations. Some of them possess next quality: absence of sterility, high content of grains on ear, high productivity, and constancy in subsequent generations, high viability.

Diversity of initial material for hybridization increased considerable. The mutants were crossed easily among oneself, with initial variety, and other varieties. Mutants with valuable characters serve as the donors for hybrid progeny. Because of this reason biological diversity is increased. In subsequent generations after crossings comes constancy. Transgression by crossings become stronger the row of quantitative characters. These characters take significant place by forming of harvest, increasing of viability, quality, resistance to pathogens.

A great interest take mutants with new characters absentee in winter common hexaploid wheat and rare for that wheat characters. This characters are impossibly or difficultly to obtain out of method of chemical mutagenesis. The genic analysis showed that changed complexes of characters are not be conditioned by pleiotropic effect, but are the result of plural mutations. As a result some correlations between characters come weaker particularly the connections which unite the desirable and undesirable characters. Therefore arise new complexes valuable characters, including new and rare characters. As a rule such complexes not arise out of method of chemical mutagenesis. All above-said explains the expansion of possibilities of selection by application of the most effective chemical mutagene in particular revival of intraspecific (intervarietal) hybridization. The latest is very effective in contradistinction of distance hybridization in the presence of extensive genotypic and phenotypic diversity in assortment of initial material and valuable characters, which fill absentee in our species of wheat characters and the new complexes of valuable characters, which are heritable in hybrid progeny.

By the method of intraspecific (intervarietal) hybridization of our mutant, as the donor of character of resistance powdery mildew, root rots, stinking smut, resistance against cold, and drought-resisting, the variety Imeni Rapoporta was created by crossing with variety Mironovskaya 808.This variety was induced in Moscow province. In farms Dmitrov's region variety Imeni Rapoporta occupies the area under cultivation 800 hectares. The farm Rogachyovo gathers in the harvest 45 quintal on 1 hectare on the area 150 hectares. In unfavourable 2003 they gather in the harvest of this variety 38 quintal on 1 hectare.

In: Biotechnology, Agriculture and the Food Industry
Editor: G. E. Zaikov, pp. 133-139
ISBN 1-60021-040-6

Chapter 22

FEATURES MICROCLONAL REPRODUCTION OF O-TYPES OF SUGAR BEET (BETA VULGARIS L.)

*O. L. Klyachenko**
National Agriculture University,
Kyiv 03041, Ukraine

ABSTRACT

The opportunity of mass reception, duplication and preservation of genetically identical initial material was investigated. As Ferilizer O-type were used as donors. Regeneration of sprouts was induced two ways: direct regeneration from explants and by formation morphogenestic structures, and then microsockets from callus.

Key words: *sugar beet, O-types, callus formation, morphogenesis, risogenesis.*

Sugar beet (Beta vulgaris L.) concerns to the major agricultural crops, being the basic producer of sugar in zones of a temperate climate. Methods of her selection constantly demand development and perfection of new effective methods of duplication of sugar beet at various stages ontogenesis with the purpose of creation of new highly productive forms of plants.

One of the basic problems of conducting selection of sugar beet on a basis cytoplasmic man's sterility still has process of creation and maintenance of selection-valuable lines - fixers of sterility. As domestic and foreign practice shows, in most cases apply self-sterile and self-fertile lines of O-type, and also selected among homozygous inbreeding lines [1,5,6]. Among domestic grades polyspermous and single-seed sugar beet self-fertile forms meet rather seldom. For this reason their search is connected to the big amount of works on self-pollination of plants.

However self-pollination and other receptions do not give a guarantee of full safety of a genotype of plants. Ferilizer O-type meet extremely seldom and the majority of them appear

* National Agriculture University, 15, Heroev Oborony St., Kyiv 03041, Ukraine. E-mail: Klyachenko@ukrpost.net

specific for certain MS forms. Maintenance of lines of sugar beet with high ability to fix CMS is closely connected to their pollen and seed efficiency [6]. Therefore most expediently culture of tissues as a method of vegetative preservation of a genotype in cleanliness [1] that is actual for practical selection.

Cultivated in vitro callus tissues of plants are characterized by a high degree of heterogeneity, even when they are received from the same donor genotypes, explants and in identical conditions cultivation. Researchers mark communication of callus morphology with their ability to regeneration [7-9]. Display of reclaiming ability explants sugar beet essentially is influenced by structure of nutrient mediums and concentration of regulators of growth [3]. Numerous experiences confirm, that at an induction of reclaiming processes on various fabrics of beet there are significant difficulties. They are caused by genetic features, a parity and concentration regulation substances [2].

The purpose of our work was research of laws callus- and morphogenesis at O-types of sugar beet and the factors influencing this process.

Technique

As donors used sprouts of six various O-types of sugar beet. Researches were carried out in several directions: ability of sprouts of sugar beet to callus formation, and then microsockets from callus was studied, and also intensity of formation sprouts is direct from explants and their rooting. Callus was induced from sprouts on Murasige- Scoog medium added 0,2-0,3 mg/l 6-BAP, 1-2 mg/l IAA, 1-2 mg/l IBA, 100 mg μ-inosit and 500 mg hydrolysate casein. Callus was cultivated in diffused light at temperature +25°C. At the end of zero passage (for 25-30 day) callus cells considered under a microscope. Regeneration of plants from callus tissues was carried out on Murasige- Scoog medium added 0,1 mg/l NAA, 0,02 mg/l 6-BAP, 0,3-1 mg/l IBA.

For studying direct regeneration of sprouts as explant was used petioles which cultivated on Murasige-Scoog medium added 0,1-0,3 mg/l IAA, 0,5-0,8 mg/l 6-BAP, 1-2 mg/l GA, 10 mg/l gljutamin. An experimental material was raised at temperature +25-26°C, illumanation 4-5 klk and 16-hour photoperiod. Rooting of plants-regenerates carried out on medium with 1-2 mg/l IBA; 0,3-0,5 mg/l GA; 0,5 mg/l NAA.

Results of Researches

At cultivation explants on a surface of a nutrient medium during first three days was observed increase initial explant. The carried out researches have shown, that in samples of O-types of sugar beet on nutrient mediums is formed callus tissue, since 7-12 day cultivation. The best nutrient medium for callus formation is Murasige-Scoog medium containing 0,3 mg/l 6-BAP and 1,5 mg/l IBA (tab. 1). Callus on her was formed dense, greenish color and the maximal weight 1,15 g.

Table 1. Dependence callus formation from the contents of regulators of growth into nutrient medium

Donor genotype	N Variants	Contens of a nutrient medium	Quantity explants, p	Quantity explants, formed callus		Character of callus
				%	Weight of callus, mg	
OT1	1	MS+0,2 mg/l 6-BAP +1 mg/l NAA	50	100	52	Light, friable dense with green sites
	2	MS+0,2 mg/l 6-BAP +1 mg/l IBA	50	100	82	
	3	MS+0,3 mg/l 6-BAP +1,5 mg/l IBA	50	100	95	
OT 2	1	MS+0,2 mg/l 6-BAP +1 mg/l NAA	50	100	40	Light, friable dense with green sites
	2	MS+0,2 mg/l 6-BAP +1 mg/l IBA	50	100	58	
	3	MS+0,3 mg/l 6-BAP +1,5 mg/l IBA	50	100	98	
OT 3	1	MS+0,2 mg/l 6-BAP +1 mg/l NAA	50	100	52	Light, friable dense with green sites
	2	MS+0,2 mg/l 6-BAP +1 mg/l IBA	50	100	75	
	3	MS+0,3 mg/l 6-BAP +1,5 mg/l IBA	50	100	1,1	
OT 4	1	MS+0,2 mg/l 6-BAP +1 mg/l NAA	50	100	53	Light, friable dense with green sites
	2	MS+0,2 mg/l 6-BAP +1 mg/l IBA	50	100	77	
	3	MS+0,3 mg/l 6-BAP +1,5 mg/l IBA	50	100	1,15	
OT 5	1	MS+0,2 mg/l 6-BAP +1 mg/l NAA	50	100	48	Light, friable dense with green sites
	2	MS+0,2 mg/l 6-BAP +1 mg/l IBA	50	100	68	
	3	MS+0,3 mg/l 6-BAP +1,5 mg/l IBA	50	100	1,01	
OT 6	1	MS+0,2 mg/l 6-BAP +1 mg/l NAA	50	100	53	Light, friable dense with green sites
	2	MS+0,2 mg/l 6-BAP +1 mg/l IBA	50	100	77	
	3	MS+0,3 mg/l 6-BAP +1,5 mg/l IBA	50	100	1,05	
HCP_{05}					19	

The important limiting factor by development of methods of reception of new forms of plants in tissue culture is ability of callus to an induction morphogenesis. The best reclaiming ability have dense dim callus greenish color which consist from roundish fine cells with rich cytoplasm and the big nucleus [4]. This description is typical for meristemoide cells which realize unique property of a plant cell - totipotency. At cultivation the given type callus on the regeneration Murasige-Scoog medium added 0,1 mg/l NAA, 0,02 mg/l 6-BAP, 0,3-1 mg/l

IBA in enviroment of intensive illumination there is his differentiation. Regeneration begins with formation of a point of growth (that is groups of cells which actively share) from which further there was a development of sprouts (fig 1, 2). Thus formation completely generated microsockets in height up to 15 mm were observed for 4-5-week cultivation (tab. 2). Further them used for rooting and reception of high-grade plants-regenerates.

Fig 1. Induction of morphogenetic callus of sugar beet O-type in vitro

Fig 2. Induction of plants regeneration of sugar beet O-type from morphogenetic callus

Table 2. Regeneration of sprouts from callus lines of O-types of sugar beet

Donor genotype	Quantity planded explants, p	Regeneration of sprouts from callus, %	Average length of sprouts, mm	Average of sprouts, p
OT 1	50	36	10	20
OT 2	50	24	8	18
OT 3	50	28	8	18
OT 4	50	36	15	20
OT 5	50	32	14	21
OT 6	50	30	15	18
HCP_{05}		5	3	7

From references it is known, that intensive regeneration of sprouts is observed from own fabrics of stem segments, petioles, and also at wound central rib a leaves [2]. For studying direct regeneration of sprouts as explant was used petioles, having high organogenic ability which was cultivated on Murasige-Scoog medium added 0,1-0,3 mg/l IAA, 0,5-0,8 mg/l 6-BAP, 1-2 mg/l GA, 10 mg/l gljutamin. The nutrient medium for formation sprouts contains substance of group cytokinins 6-BAP in concentration 0,5-0,8 mg/l which causes intensive formation of kidneys (fig. 3). Numerous sprouts and kidneys are formed during two weeks. In 3-6 weeks cultivation was carried out the account of formation sprouts (tab. 3).

The important stage in reception meristeme the plants ready to landing in a ground, is process of rooting developed sprouts. The sprouts of one size with well-developed leaves were selected for rooting and planded into nutrient medium. At selection of an optimum nutrient medium for rooting sprouts were tested medium with ½ concentration macro-, microelements, vitamins and added with regulators of growth. One authors [3] recommend to apply to rooting the diluted mineral medium without regulators of growth, others [2,7] advise to add auxin. Analyzing results of researches submitted in tab. 4 it is possible to note, that medium with addition 1,5 mg/l IBA (fig. 4), which absence slows down root system especially in an initial stage of development is optimum. About 95 % explants on the given environment had well advanced root system.

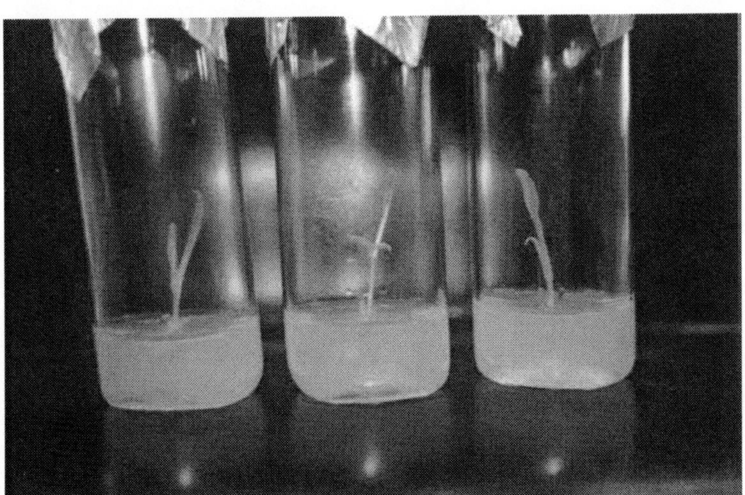

Fig 3. Induction of a direct morphgenesis from explants of sugar beet O-type

Table 3. Formation sprouts O-types of sugar beet from explants

Time of the account	Donor genotype	Sprouts in length 1-2,9 sm	Sprouts in length 3-4,9 sm	Sprouts in length of 5 sm and are higher	In total sprouts
Quantity sprouts in 3 weeks cultivation, p	OT 1	32	16	2	50
	OT 2	30	15	3	48
	OT 3	37	13	4	54
	OT 4	32	14	4	50
	OT 5	37	18	5	60
	OT 6	30	16	2	48
Quantity sprouts in 6 weeks cultivation, p	OT 1	43	22	8	73
	OT 2	45	21	7	73
	OT 3	49	28	5	82
	OT 4	44	22	6	72
	OT 5	50	29	11	90
	OT 6	52	24	12	88
HCP_{05}					

Table 4. Influence of structure of a nutrient medium on rooting of runaways of O-types of sugar beet

Variant of medium	Donor genotype	Quantity planded sprouts, p	Quantity of rooted sprouts,	
			p	%
MS + 0,5 mg/l GA + 0,5 mg/l NAA	OT 1	50	26	52
	OT 2	50	28	56
	OT 3	50	26	52
	OT 4	50	24	48
	OT 5	50	23	46
	OT 6	50	28	56
MS+0,5 mg/l NAA	OT 1	50	10	20
	OT 2	50	13	26
	OT 3	50	16	32
	OT 4	50	12	24
	OT 5	50	18	36
	OT 6	50	15	30
MS+1,5 mg/l IBA	OT 1	50	47	94
	OT 2	50	48	96
	OT 3	50	47	94
	OT 4	50	48	96
	OT 5	50	48	96
	OT 6	50	46	92
HCP_{05}				

Fig 4. Rootage of plants-regenerates of sugar beet O-type

In 2-3 weeks cultivation plants-regenerates in test tubes of them were replaced in an easy substratum preliminary having washed root system from the rests of medium. At adaptation of plants it is important to maintain constant temperature, humidity and illumination. Strice-root regenerates in a substratum has made 94 %. Acclimatized plants of O-ypes of sugar beet in 3-4 weeks were replaced in an open ground.

Thus, we have developed a number(line) of the biotechnological receptions, allowing to receive a plenty of valuable genotypes. Optimum enviroments of cultivation and acclimatization of plants-regenerates are picked up.

REFERENCES

[1] I.J. Balkov: *CMS sugar beet.* Kiev, 1990. 239 p. (in Russian).

[2] M.A.Bannikova, A.E.Golovko, O.A.Hvedynych, N.V.Kuchuk: *Cytology and genetics.*, (6), 14 (1995).

[3] F.L.Kalinin, G.P.Kushnir, V.V.Sarnatskaja: *Technology microclonal reproduction of plants.* Kiev, 1992. 154 p. (in Russian).

[4] O.L. Klyachenko: *Agrarian science and education.*, 4 (3-4), 24 (2003).

[5] I.I. Ljal'ko, I.A. Shevtsov, F.N. Pary: *Cytology and genetics.*, 24 (2), 41 (1990).

[6] V.P. Oshevnev, N.P. Gribanova: *Reports of the Russian academy of agricultural sciences.*, (1), 11 (2003).

[7] G.N. Jurkova, O.V. Dubrovnaja, T.V. Chugunkova, I.I. Ljal'ko: *Physiology and biochemistry of cultural plants.*, 31 (3), 227 (1999).

[8] W.P Doley: *Plant Cell Rep.*, (8), 222 (1989).

[9] F.A. Krens, D. Jamar: *J. Plant Physiol.*, (134), 651 (1989).

In: Biotechnology, Agriculture and the Food Industry
Editor: G. E. Zaikov, pp. 141-144

ISBN 1-60021-040-6

Chapter 23

Polymorphism of 10 Microsatellites and Their Usefulness for Paternity Control in Dogs

Anna Radko, Ewa Słota and Mariusz Kościelny

Department of Immuno and Cytogenetics,
National Research Institute of Animal Production,
ul. Krakowska 1, 32-083 Balice, Poland

Abstract

The aim of the present study was the evaluation of microsatellite DNA markers for paternity control in dogs using automated DNA sizing technolog. Analysis of microsatellite polymorphism used markers PEZ1, PEZ3, PEZ5, PEZ6, PEZ8, PEZ11, PE12, PEZ18, UCB2054, UCB2079 from the "CANINE PCR MARKER KITS for pedigree verification" kit (*Applied Biosystems*). The polymorphism of a total of ten microsatellite loci were analysed in 14 individuals of Barzoi dog. Blood samples were used to isolate genomic DNA. Isolated DNA was used to amplify microsatellite loci by polymerase chain reaction (PCR).

PCR products were separated electrophoretically on a 4% denaturing polyacrylamide gel in laser sequencer ABI PRISM 377 for vertical electrophoresis.

The results of electrophoretic separation were analysed automatically with GeneScan 2.1 software, and the genotypes were established by Genotyper 2.0 software.

The frequency of identified alleles was used to compute the degree of heterozygosity (H), the polymorphic information content (PIC) and probability of exclusion (PE) for each microsatellite marker studied.

Key words: *dog, microsatellite DNA, parentage control*

Introduction

In animal breeding accurate determination of relatedness and efficient control of pedigree registration is of great importance. Traditionally, pedigree verification in cattle, sheep, pig, horses and dogs has been carried out using blood groups and protein polymorphisms. However, the number of loci used gives relatively low precision and cannot always exclude a putative sire, even if that putative sire is not the real parent. Molecular techniques using DNA markers are more specific than the previously used technique of blood typing. Additionally, blood typing test can only be performed with fresh blood, while the DNA test can be performed with different samples (blood, hair, semen or saliva). DNA-based technologies enable detection of different polymorphic types. Among these, microsatellites or short tandem repeats (STRs) have been identified in all eukaryotic species so far investigated. Due to their relative abundance in the genomes and high polymorphism they are important tools in pedigree verification in domestic animals [1,2,5,6,7].

The objective of this study was to analyse the polymorphism of ten microsatellite DNA markers and their usefulness for parenting verification in a Barzoi in Poland.

Material and Metods

Polymorphism of 10 microsatellite loci was analysed in 14 blood samples from Barzoi breed in Poland.

DNA polymorphism was analysed using microsatellite markers: PEZ1, PEZ3, PEZ5, PEZ6, PEZ8, PEZ11, PE12, PEZ18, UCB2054, UCB2079.

Genomic DNA was isolated from 40 μl of peripheral blood using proteinase K. Based on the isolated DNA, the microsatellites were amplified by the polymerase chain reaction (PCR). Standard reagents "CANINE PCR MARKER KITS for pedigree verification" by *Applied Biosystems* were used for PCR. PCR products were separated by electrophoresis in a 4% denaturing polyacrylamide gel on laser sequencer ABI PRISM 377 (Applied Biosystems). Results of the electrophoretic separation were analysed with GeneScan 2.1 software and the alleles were sized with Genotyper 2.0 software. Based on the frequency of the identified alleles, the degree of heterozygosity (H), the polymorphism information content (PIC) and the probability of exclusion (PE) were calculated for each microsatellite marker studied. PE was calculated with regard to the possibility of examining both parents. The combined probability of exclusion (PEc) was calculate for 10 analysed markers.

Results And Discussion

The great number of DNA microsatellite sequences identified so far, characterized by a high degree of polymorphism and the development of automated methods for their analysis made it possible to investigate the genetic variability and pedigree verification in dogs based on microsatellite markers.

We selected ten microsatellite markers and examined their application for parentage testing in Barzoi. A total of 45 alleles were identified for the 10 microsatellite sequences studied. The number of alleles per locus varied from 3 to 6 alleles (Table 1).

Table 1. Number of identified alleles, size range in base pairs, coefficient of heterozygosity (H), polymorphic information content (PIC) and probability of exclusion (PE) for the DNA microsatellite markers studied

Marker	No. of alleles	Size range (bp)	H	PIC	PE
PEZ1	5	110-134	0.610	0.566	0.378
PEZ3	3	119-139	0.521	0.465	0.276
PEZ5	4	105-117	0.717	0.664	0.459
PEZ6	6	182-198	0.778	0.745	0.571
PEZ8	5	223-239	0.668	0.606	0.404
PEZ11	6	130-164	0.786	0.752	0.578
PE12	4	276-294	0.630	0.567	0.366
PEZ18	3	226-238	0.571	0.501	0.301
UCB2054	5	154-174	0.622	0.550	0.351
UCB2079	4	271-291	0.365	0.343	0.201

Based on the identified alleles, the experimental material was analysed for polymorphism of genetic markers used for parentage testing in dogs by determining the polymorphism information content (PIC) and the degree of heterozygosity (H). Based on the calculated parameters of PIC and H it was found that the microsatellite markers show a high degree of polymorphism in the material studied (Tab. 1). Except the loci PEZ3 and UCB2079, which showed the lowest variation (PIC=0.465, H=0.521 and PIC=0.343, H=0.365), the calculated PIC values exceeded 0.5 and H values ranged from 57 to as much as 78%. The highest polymorphism was displayed by the locus PEZ11and PEZ6 (PIC=0.752, H=0.786 and PIC=0.745, H=0.778).

The usefulness of polymorphic genetic markers for parentage verification was determined by estimating the probability of incorrect pedigree (PE). The high probability of exclusion (PE_C=0.999), calculated based on microsatellite DNA markers was also shown for other breeds of dogs [3,4]. The probability of exclusion, calculated separately for each locus with regard to the possibility of examining both parents, is given in Tab. 1. PE values obtained for each marker served as a basis for calculating total probability of exclusion (PE_C). Based on PE_C it was found that incorrect pedigree can be excluded with 99.4% accuracy using DNA analysis.

The results suggest that the ten markers can be applied for routine parentage testing in studied dogs.

REFERENCES

[1] A.T. Bowling, M.L. Eggleston-Stott, G. Byrns, R.S. Clark, S. Dileanis, E. Wictum: *Anim. Genet.* 28 (4), 247-252p (1997).

[2] S. Denise, E. Johnston, J. Halverson, K. Marshall, D. Rosenfeld, S. Mckenna, T. Sharp, J. Edwards: *Anim. Genet*. 35 (1),14-17p (2004).

[3] M. Fredholm, A.K. Wintero: *Anim. Genet.* 1996, 27(1), 19-23p.

[4] Y. Ichikawa, K. Takagi, S. Tsumagari, K. Ishihama, M. Morita, M. Kanemaki, M. Takeishi, H. Takahashi: *Vet Med Sci.* 63 (11), 1209-13p (2001).

[5] A. Radko, M. Duniec, T. Ząbek, A. Janik, M. Natonek: *Med. Wet.* 58, 708-710p (2002). (In Polish)

[6] T. Rychlik, A. Radko, M. Duniec: *Med. Wet.* 59, 1016-1018p (2003) (In Polish).

[7] I Zajc, J. Sampson: *Pflugers Arch.* 431 (6 Suppl 2), R201-2p (1996).

In: Biotechnology, Agriculture and the Food Industry
Editor: G. E. Zaikov, pp. 145-155
ISBN 1-60021-040-6

Chapter 24

THE USE OF *MYCOBACTERIUM NEOAURUM* CELL CULTURE FOR PRODUCTION OF ANDROSTENEDIONE AND ANDROSTADIENEDIONE FROM STEROLS

N. V. Rodina, M. A. Molchanova, N. E. Voishvillo, V. A. Andryushina*, T. S. Stytsenko, T. S. Savinova and K. G. Skryabin
Centre 'BIOENGINEERING' RAS. Moscow, 117312, Russian Federation

ABSTRACT

Dependence of sterol side chain cleavage velocity as well as structure and yield of reaction products from fermentation conditions and sterol structure was investigated during sterol oxidation by the *Mycobacterium neoaurum* culture. Conditions of sterol transformation with selective mutant strains *Mycobacterium neoaurum* into AD and ADD were optimized in accordance with of the revealed characteristics. The optimum type of sorbent was found, allowing to remove bacteriastatic action of ADD onto the producing strain.

Key Words: *androstendione, androstadienedione, cholesterol, phytosterols, transformation, mycobacteria, sorbents.*

At present the main raw materials for production of steroid medicines are sterols of plant and animal origin (phytosterols, cholesterol). Cholesterol (I) is an individual compound and can be obtained from wool grease and bone marrow of cattle and sheep. Phytosterols (PS), usually representing sterol mixture of stygmastan and ergostan series, are preferable as a substrate for steroid synthesis, as they are waste materials from wood manufacture and processing of soybean and sugarcane. The following sterols are prevailing in plant sources: β-sitosterol (II) , stygmasterol, campesterol, brassicasterol, ergosterol , fucosterol. Quality and quantity composition of PS depends significantly on there origin. The intermediates in the

* Centre 'BIOENGINEERING' RAS. Prosp. 60let Oktyabrya 7, kor. 1, Moscow, 117312, Russian Federation , E-Mail: Andryushina@rambler.ru Fax +007-095 1350571

synthesis of almost all range of medicinal preparations of steroid origin from sterols are 17-ketoandrostanes: androstenedione (AD), androstadienedione (ADD) and 9α-hydroxy-AD (fig. 1). These ketosteroids (KS) can be obtained by selective cleavage of side chain from sterols of plant and animal origin with the help of bacterial mutant strains, mainly mycobacterium strains, in which the synthesis of 1,2-dehydrogenase and 9α-hydroxylase (AD) is blocked, or only of 9α-hydroxilase (ADD), or only of 1,2-dehydrogenase (9α-hydroxy-AD).

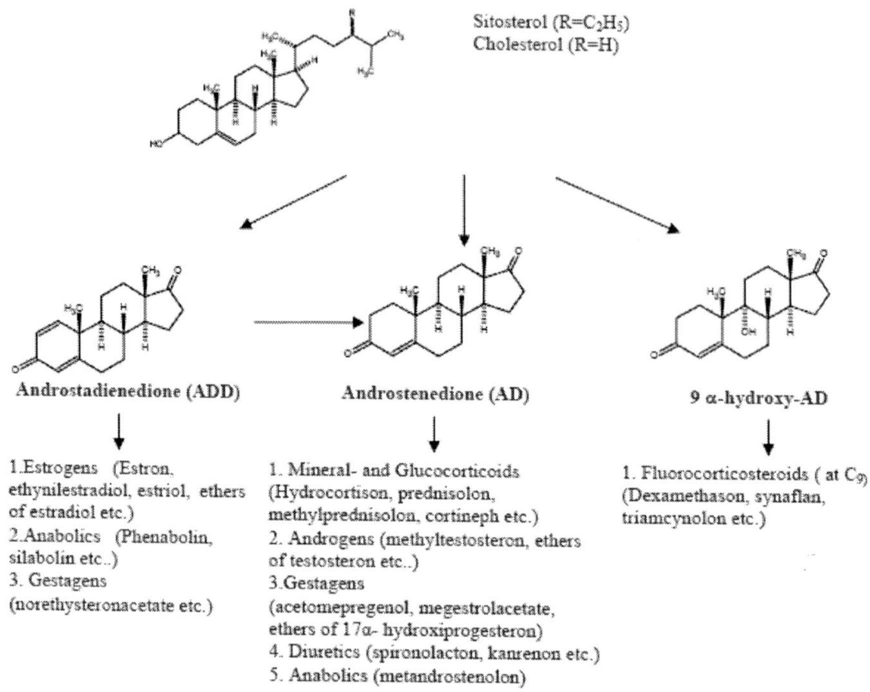

Fig. 1. Medicinal preparations produced from sterols, *via* diketosteroids

Significance of indicated KS - AD, ADD, 9 α-hydroxy-AD - as precursors in the synthesis of vitally important steroid preparations was already accentuated in the first reviews, dedicated to the problem of the sterol side chain bacterial cleavage [1-4].

It was estimated by comparison of various bacterial species activity: - *Arthrobacter, Mycobacterium, Nocardia* - towards various steroid substrates, that both the rate of production and the output of the side chain cleavage products depended on both the conditions of fermentation, and of the steroid molecule structure, namely – of the side chain length, of the branching at C-24 , of 22,23-double bond presence etc. Summarizing the data [5-7], all investigated steroids can be distributed in accordance with the diminution of the output of the side chain cleavage products – C-19- and C-22-steroid intermediates - in the following succession: pregnenolon>> lythocholic acid> cholesterol >campestrol, β-sitosterol >stigmasterol > 7-dehydrocholesterol > ergosterol, fucosterol. In a series of sterols, I is at the first place. Among phytosterols, for KS production less preferable are the sterols with double bond in the side chain (stigmasterol, fucosterol). It is obvious that the rate of steroid penetration into the cell is determined by the degree of their hydrophoby. In the work of Sato et al. [8],. the direct dependence of the sterol transport into the cell from the side chain structure was demonstrated.

The analysis of data from scientific papers and patents [7,9-11], allows to discuss the main causes, decreasing the amount and the speed of formation of the products of sterol side chain cleavage at conversion from I to PS. In the fundamental paper of Dr.Sih C.J. [9-11], awarded with the prize «Le Prix Roussel», it was estimated that as against of I, for cleavage of the side chain of sterols with branched side chains at C-24, the presence of HCO_3^- ion is necessary, - its introduction into C-28 position preceded cleavage of C-24-C-25- and C-24-C-28-bonds in PS. Enzymatic introduction of HCO_3 ion into steroid molecules was proved by 3-keto-24-ethylcholest-4-ene-26-acid incubation with cell-free extract of AD-producing bacteria *Mycobacterium sp*. B-3805 in the presence of labeled with CO_2, ATP and coenzyme A [11].

Addition of bicarbonate ion into the medium increased the rate of formation and output of KS during incubation of II and campestrol with AD- and ADD-producing strains of *Mycobacterium sp* B-3605, B-3683 and *M.phlei* NRRL B-15050 [10,11].

The second factor, determining differences in bacterial transforming activity towards I and II, is the difference in the mechanism of bacterial consumption of these substances. [7]. It was demonstrated that water-insoluble compounds of steroid kind can be consumed by microorganisms partly in solubilized form, but mainly – as a result of absorption/adsorption of sterol molecules in the area of substrate-cell contact, appeared due to cell adhesion to sterol particles. The presence of synthetic surface active compounds, preventing cell adhesion to II particles, lowered significantly the level of bacterial consumption of sterols [7]. The presence of the direct contact of microbial cell with substrate crystals was shown in the work of Atrat et al. [12].

It was experimentally estimated, that II, as more hydrophobic substrate comparing with I, is consumed by bacteria principally as a result of adsorption, whereas I approximately for 30% is consumed by cells in solubilized form [7].

The aim of the present work was, taking into account all mentioned above data, to optimize the conditions of transformation for soybean PS, with the help of mutant strains of bacteria *Mycobacterium neoaurum*, capable to cleave selectively the side chain of I and PS, without degradation of steroid nucleous, with production of AD or ADD.

For this purpose the following was fulfilled:

1) Study of the influence of soybean meal lipids onto conversion of I and PS into AD with *Mycobacterium neoaurum* VCIM Ac-1634 cell culture .
2) Study of the dynamics of AD accumulation during I and PS, depending on the type of reaction mass mixing and on the glucose concentration in transformation medium.. Determination of the optimum amount of glucose.
3) Study of the influence of the inoculate amount of *M. neoaurum* VCIM Ac-1634 onto the velocity of the AD fermentation product formation. Study of the dynamics of the fermentation velocity, depending on the cell concentration of the strain, transforming substrate.
4) Determination of the bacteriastatic concentration of ADD at I or PS transformation with *Mycobacterium neoaurum* VCIM Ac-1656 culture.
5) Determination of dependence of ADD formation during I or PS transformation with *M. neoaurum* VCIM Ac-1656 culture, on modification and amount of sorbent

«Porolas». The choice of the sorbent and determination of optimum conditions for transformation.

MATERIALS AND METHODS

For sterol conversion into AD and ADD, the bacterial mutant strains VCIM Ac-1634 and VCIM Ac-1656, correspondingly, obtained from the parent strain of *Mycobacterium neoaurum* [13], and differing from the parental strain by the ability to cleave selectively sterol side chain with formation of AD and ADD.

Cell cultures were maintained and grown in conditions, described earlier [14]. Medium for transformation was prepared with demineralized water and differed from the used earlier [14] by glucose content (1-4%) and soybean flour composition (Tab.1), pH before sterilization 7,0-7,3. Sterols were introduced into medium before sterilization as water suspension, containing 0,3-0,5% of Tween-80. PS and I were disintegrated in ball mill up to particle size not exceeding 10 μ. Substrates for transformation were cholesterol of 98% purity and a mixture of soybean PS (25,6% II, 23% campesterol, 22,3% stigmasterol and 12,6% of unidentified sterols). For transformation of sterols into ADD, synthetic non-ionogenic polymer of «Porolas» type was introduced into medium before sterilization, in ratio to sterol amount 8-10/1,0. The medium with I was poured by 100 ml into conical flasks of 750 ml, The medium with PS – into flasks of the same shape, but supplied with baffles. Inoculate for sterol transformation into AD was introduced into the medium in amount of 10-30 vol..%; for transformation into ADD – 10 vol.%. Transformation was carried out at 30° C and mixing on the shaker (220 rpm). The CFU value and dry weight were determined in the inoculate and in samples, taken during transformation. Glucose content was determined by glucosoxidase method.

The optical density was measured at wave length 480-520 nm. AD content in samples, as well as AD or ADD concentration in cultural liquid were determined after the end of transformation with the help of TLC and HPLC. Concentration of PS residual amount was detected by semi-quantity method of TLC [13]. For TLC control, plates «Silufol UV-254" (Czech Republic) were used. HPLC determination was carried out on Gilson device, on column with Silasorb 18 (4,0x250 mm), having grains 10μ, λ 254 nm, at room temperature.

Table 1. Quality and quantity composition of soybean flour samples

Soybean flour sample	Color	Particle size, μ	Composition (%)			
			Fat	Protein	Ashes	Moisture
Degreased-1	Light-yellow	200	0,96	47,88	10,4	7,71
Non-degreased -2	Yellow	200	17,79	50,61	6,5	6,79

The sample was extracted with ethylacetate (1:5*)*. The extract obtained was evaporated till dry residue, and dissolved in 60-65% acetonitril, eluated also with water acetonitril. The flow velocity is 0,8ml/min, mobile phase MeOH-H_2O (7:3). The time of ADD output is– 5,8 min, of AD- 8,0 min. Extraction of transformation product was carried out according to [14]. Molar output of AD or ADD, using PS as a substrate, was calculated, assuming average value of the data of quantitative analysis PS. Amount of PS loaded was determined, taking into account transformed sterols in an initial substratum totally.

Results and Discussion

1). Study of the influence of lipids amount in soybean flour onto conversion of I and PS in AD.

In a number of works [12,15,16] there was shown the positive influence of fatty acids onto the output of sterol side chain cleavage products. Particularly, in a work with nokardia [16] it was noted that the oxidation of β-[4-C^{14}-]-sitosterol increased in the presence of palmitic or stearic acid. It was assumed that fatty acids are used in biosynthesis of surface-active lipids of cell wall, that raises its permeability for lipophylic compounds.

The impact of lipid content onto I and PS transformation into AD with *M.neoaurum* was studied in the present work in medium, prepared with soybean flour of various fattiness (Table1).

The AD output from I in the medium with non-degreased flour was 36% higher than in the medium with degreased flour (Table 2), whereas the difference in AD output from PS was not significant (13%). It was not possible to reach complete PS conversion by elongation of the transformation time. If the PS fermentation time increased, then cell agglomerates accumulation occurred, as well as amount of reaction side-products was increased.

Table 2. AD accumulation, from I and PS, transformed with *M.neoaurum* VCIM Ac-1634 in mediums with soybean flour of various fattiness in presence of 1% glucose

Substrate, 10 g/l	Soybean flour*	Transformation time, hours			
		15	30	50	75
		AD, g/l			
Cholesterol	1	0,1	1,2	2,2	3,1
Cholesterol	2	0,6	2,5	3,7	5,8
Phytosterol	1	-	0,4	1,2	1,7
Phytosterol	2	0,05	0,7	2,1	2,5

* 1-degreased (fattiness 0,96 %); 2-non-degreased (fattiness 17,79 %)

2). Influence of the mixing type of cultural liquid and glucose concentration onto the transformation of PS into AD process.

As the Table 2 shows the product yield from side chain cleavage of I is higher, comparing with II., at sterol transformation with *M. neoaurum* at the same conditions. Maximum AD yield from I was 78%, from PS – only 40%.

Analogues side chain cleavage of I and 3β-acetyl-I with *Mycobacterium sp.* M12 culture took place with accumulation of 69-70% AD, whereas at transformation of II and 3β-acetyl-II, the yield of AD was only 22-26% [15]. Low yield of AD from PS was observed at laminar mixing in standard shaker flasks. However, at fermentation in flasks with baffles, the amount and rate of AD formation from PS increased (Fig. 2, curves 1,2). Our explanation of this phenomenon is, that in flasks with baffles the laminar type of mixing changed for turbulent one. Whirlwind flows in reaction mixture prevented from agglomeration of cells and substrate particles, and, most important, increased saturation of cultural liquid with carbon dioxide, excreted by bacteria.

It is well known, that at certain concentration of calcium ions in water medium and dissolved carbon dioxide, the following reaction takes place: $CaCO_3 + CO_2 + H_2O \Leftrightarrow Ca^{2+} + 2HCO_3^{-}$.

Turbulent flows provide sufficient amount of CO_2 in cultural liquid, and reaction balance moves towards formation of HCO_3^- ions, because they are spent during the process of PS side chain cleavage. For providing of reaction system with hydrocarbonate ion, 2,5 g/l of $CaCO_3$ was added into medium for fermentation of 10 g/l PS.. Amount of seeding material – 10 vol.% was the same, as for I transformation. At laminal character of mixing, taking place during fermentation in standard shaker flasks, intense formation of HCO_3^- ion didn't occur, because heavy particles of calcium carbonate were localized mainly in bottom lay of cultural liquid.

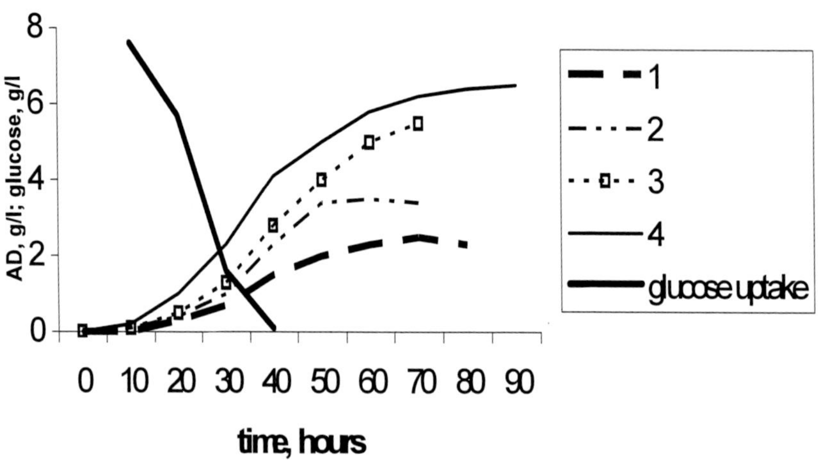

Explanation to Fig.2.

Curve	Inoculum, g/l	Glucose, g/l	Mixing
1	0,15	10,0	Without baffles
2	0,15	10,0	With baffles
3	0,15	20,0	With baffles
4	0,45	40,0	With Baffles

Fig. 2. Influence of the transformation conditions on the time course of AD production

However, in experiments with PS, glucose concentration in medium was important. At initial glucose concentration in medium 1% and the use of 10% inoculate, the reserve of glucose was already exhausted after 40 hours of fermentation, and PS transformation stopped with AD production 3,5 g/l. The result of growth of initial glucose content in medium up to 2% was complete PS conversion into AD (Fig. 2, curves 2,3).

3) Study of the dependence of AD accumulation rate from the amount of M. neoaurum VCIM Ac-1634 seeding material.

We observed that AD already appears in the medium before the beginning of cell number doubling of *M. neoaurum VCIM* Ac-1634. Earlier the same phenomenon was revealed [6] in experiments with *Arhrobacter sp*. We assumed that the rate of AD production and its amount depend on the quantity of seeding material.

To examine the dependence of the rate of AD formation at PS transformation with *M. neoaurum VCIM* Ac-1634 culture from the amount of inoculum, we carried out the experiment with the seeding material dose, 3 times enlarged (30 об.%). The optimum glucose amount, determined experimentally for the indicated dose of seeding material, 4%. During first 40 hours of incubation (i.e. in the period of mycobacteria intense growth) the rate of AD formation in the experiment with triple amount of inoculum was a little higher - from 20 to 180 mg AD/l/h (10-40 hours from the beginning of fermentation with 30%-content of inoculum), comparing with the control - from 10 to 150 mg AD/l/h (10 vol..%). On the Fig. 3 one can see, that the rate of AD accumulation with 30% inoculum is twice higher only during first 30 hours of incubation. Then the lowering of the fermentation velocity takes place, and in the stationary phase of growth it is rather lower than, the rate of AD production in the control. In spite of the fact, that at increasing of the seeding material amount 3 times the rate of the transformation product accumulation and its amount rose only during first 40 hours of fermentation, nevertheless this approach enabled to increase the substrate amount in the medium and reach high degree of PS conversion into AD (Fig.2, curve 4).

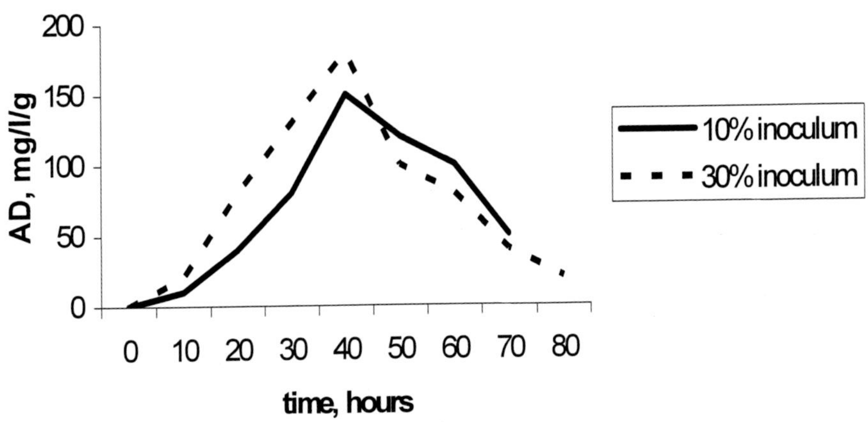

Fig. 3. Rate of AD accumulation with different amounts of inoculum

To increase the rate of production and amount of KS forming by bacteria *Arthrobacter sp*. 317 from PS, the density of seeding material was enlarged 4 times comparing with inoculum for transformation of the same amount of I (control). However, at laminar mixing, the content of KS increased not more than 1,5 times, and the velocity of their production stayed the same, i.e. 3 times lower than control value[7].

It was earlier reported [17], that at glucose concentration above 1,5%, the main products of sterol side chain cleavage, were the products of 17-ketogroup reduction - testosterone (TS) and 1-dehydro-TS. However, in our case, at rather high PS conversion 85-90%, AD was the major transformation product, with admixture of insignificant amount of side-products TS, 1-dehydro-TS and hydroxymethylpregnan (HMP) (Table 3 – average data from several operations are presented).

Table 3. The content of products of PS side chain cleavage after 90 hours of transformation depending on the dose of seeding material of *M.neoaurum* and glucose concentration in the medium

Amount of PS*, g/l	Amount of glucose, %	Biomass, g/l		Extracted steroids					
				Transformation products, %					Residua PS, g
				AD		Admixtures in the product of technical grade			
		Inoculum	At fermentation end	Product purity	Yield	TS	Δ^1-TS	HMP	
12,75	4,0	0,42	28,6	96,9	71,2	0,52	0,1	1,02	1,8
9,9	1,0	0,15	15,7	95,5	50,3	traces	traces	0,4	2,7

* Calculated for 100% content of PS in substrate

4) Determination of the bacteriastatic concentration of ADD at I or PS transformation with Mycobacterium neoaurum ВКПМ Ac-1656 culture.

In experiments with bacterial strains, capable to cleave sterols with ADD formation, it was estimated that ADD, in contrast to AD, already in insignificant concentration cause bacteriastatic effect onto producing strains. Suppression of *Mycobacterium vaccae* growth was observed in the presence of 0,8 g/l ADD [18], and *Arthrobacter globiformis* growth and sugar cane sterols conversion was ceased at ADD content 0,7 g/l [19]. We determined bacteriastatic ADD concentration for *Mycobacterium neoaurum* ВКПМ Ac-1656 strain , as 1–1,5 g/l .

Table 4. Growth of *M.neoaurum* and bioconversion of cholesterol (I) depending on the presence of synthetic ADD adsorbent

№	Transformation variant I	CFU/ml	Steroid concentration	
			I, g/l	ADD, %
1	Medium after inoculation	3 x 10^7	9,8	-
2	60 h. of fermentation without adsorbent	5 x 10^7	4,0-5,0	10-16
3	60 h. of fermentation with adsorbent	4,8 x 10^9	traces	65-67
4	Adsorbent added after 30 h. from the beginning of fermentation	4,7 x 10^8	2,5-3,0	30-35
5	Adsorbent added after 20 h. from the beginning of fermentation	2,1 x 10^9	0,5-1,0	60-65

On the Table 4 one can see, that during the process of incubation of mycobacteria with the resin, which adsorbed completely ADD, produced from 10 g/l I, cell propagation and complete substrate conversion is occurred. In the absence of adsorbent, biomass growth could not be observed, ADD content in cultural liquid after 20 hours of fermentation was less than 20%, or ~ 1,0-1,5 g/l. If at that moment the sorbent was fed into fermentation medium, than biomass growth was renewed, and ADD output became the same, as in the experiment, in which the sorbent was introduced at the beginning of fermentation. If the sorbent was introduced into cultural liquid after 30 hours from the beginning of fermentation, than mycobacteria growth was renewed, but, as it could be expected, substrate conversion was incomplete, and ADD output – low. If I was replaced for PS, analogous to AD-producing strain, the decrease of ADD yield occurred, - 60-65% from I, and 20-30% from PS. The same was noted earlier [5] at and II fermentation to ADD with *Arthrobacter simplex* culture: ADD yield at I and II transformation was equal to 57,5% and 39,2% correspondingly.

5) Study of the dependence of ADD accumulation at PS transformation with M. neoaurum ВКПМ Ac-1656 culture on modification of the sorbent «Porolas»

We chose 8 polymer samples with specific surface 500 –1000 m^2/g, pore volume 1-2,5 cm^3/g and mechanical strength exceeding 90% for examination (Table 5). Ratio between amount of resin and transformed substrate was 8-10/1. Sorbents of «Porolas» mark are produced by Russian industry and represent non-polar, or slightly polar resins – products of co-polymerization of divinylbenzen and ethylstirol in the presence of modifying additives. To chose optimum sorbent, we carried out comparative study of different Porolas modifications during the process of ADD producing from PS. It is well known that analogous resins of «Vofatite» mark are able to adsorb selectively only products of sterol transformation, and don't adsorb the substrate itself [20]. On the other hand, sorbents of Porolas mark are well known for their ability to adsorb from cultural liquid the products microbial transformation of corticosteroids [21].

Table 5. Some physical characteristics of sorbents utilized for isolation of steroids

Sorbent's name	Modifying component	Surface area (m^2/g)	Bulk weight (g/l)	Mechanical stability, (%)	Volume of pores ($sm^3|g$)	Quantity of ADD (g/l)	
						In fermentation broth	on a sorbent
Porolas A	Vinyl acetate	370	520	90	1,23	0,5	1,1
Porolas B	Vinyl acetate	600	300	90	1,56	1	1,2
Porolas C	Toluene	700	300	80	1,72	0,3	2,0
Porolas E	Methacrylic acid	850	300	95	3,10	1,5	2,3
Porolas F	Vinyl acetate	900	300	95	3,50	-	4,5
Porolas G	Vinyl acetate	1074	320	95	3,90	-	4,6
Porolas H	Vinyl acetate	1309	300	90	4,30	1,4	4,5

As the most perspective Porolas G was chosen from the investigated sorbents due to its higher mechanical stability - 95 % and high sorbtion capacity of 3.90 sm3 /g. High mechanical stability of a sorbent had priority value in connection with obligatory use of baffles in PS transformation process into ADD , as well as for AD-production. In conditions for transformation – 100 мl of adsorbent Porolas G for 10 g of sterol in 1 l of medium of the same composition as for AD production, but with 1 % glucose and 10 % of seeding material; resin was introduced simultaneously with the seeding material, - at these conditions ADD yield after 65 hours of PS transformation was 65-70%. Side products content (AD and 1-dehydro-HMP) varied in the limits of 10-12%, TS and 1-dehydro-TS were absent.

Conclusion

1) As a result of the carried out research it is established, that to obtain the maximal specific activity of AD and ADD forming strains, as well as high speed of the target products formation, complete substrate conversion and absence of by-products in reaction of 17-keto-group reducing, it is necessary to provide realization of fermentation process in a mode of intense turbulent agitation in combination with the high contents of glucose in the medium.
2) The mode of intensive turbulent mixing of the reaction medium promotes its saturation with carbon dioxide, confirmed by pH decrease from 7,0-7,3 up to 6,4-6,6 during fermentation. At laminar agitation such decrease is not observed.
3) Intense agitation prevents from agglomeration of the substrate particles and from agglomeration of cells, characteristic to these transforming bacterial strains.
4) In the offered mode of fermentation the specific activity of biomass during of ADD formation is higher, than during AD formation , apparently, due to removal of transformation product, toxic for mycobacteria.

References

[1] S.Writer: *Technocrat,* 8(3), 30 (1975).

[2] C.K.A.Martin: *Adv.Appl.Microbiol.*, 22(1), (1977).

[3] U.Schoemer, C.K.A.Martin: *Biotechnol. and Bioeng.*, 22(Suppl.1),29 (1980).

[4] K.Kieslich: *J.Basic Microbiol.*, 25(7), 461 (1985).

[5] M.Nagasawa, N.Watanabe, H.Mashida et al.: *Agr. Biol. Chem.*, 34(5), 798 (1980)

[6] P.G.Goswami, H.D.Singh, J.N.Baruah: *Folia Microbiol.*, 29, 209 (1984).

[7] P.G.Goswami, H.D.Singh, S.D.Bhagat, J.N.Baruah: *Biotechnol. and Bioeng.*, 25, 2929 (1983).

[8] Y.Sato, 8.K.Nishikawa, K.Aikawa et al.: *Biocim. Biophys. Acta*,38, 1257 (1995).

[9] Y.Fujimoto, C.S.Chen, A.S.Gopolan, C.J.Sih, et al.: *J.Am.Chem.Soc.* 104, 4720 (1983).

[10] Y.Fujimoto, C.S.Chen, Z.Szeleczky, D.D.Tullio, C.J.Sih: *J.Am.Chem.Soc.* 104, 4718 (1983).

[11] C.J.Sih: US Patent 4 444 884, Cl. 435/55 (1984).

[12] P.Atrat, P.Hosel, W.Richter, H.W.Meyer, C.horhold: *J.basic Microbiol.*, 31(6), 413 (1991).

[13] N.E.Voishvillo, V.A.Andryushina, T.S.Savinova et al: *Appl. Biochem. and Microbiol.* 39(2), 152 (2003).

[14] V.Andryushina, N.E.Voishvillo, K.G.Skryabin: Patent Ru 2231553, (2004).

[15] L.O.Zhang, E.P.Bian, Y.Wang: Yaoxue Xuebao, 27, 903. *Chem. Abstr.*, 118, 123024 (1993).
[16] U.Schoemer, F.Wagner: Eur. *J. Appl. Microbiol.*, 10(12), 99 (1980).
[17] M.A.Garabal, M.A.D.Perez, C.P.Bolanos, C.P.Fonseca: Rev. CENIC, *Cienc. Biol.*, 30(1), 17 (1999).
[18] C.Horhold, B.Gottschaldt, H.H.Grosse: Fifth Int. *Conf. on Chemistry and Biotechnol.* Varna, Bulg., 4, 92, (1989).
[19] S.K.Srivastava, R.A.K.Srivastava, S.N.Mathur: *J.Appl.Bacteriol.*, 59(5), 399 (1985).
[20] B.Gottschaldt, H.H.Grose, C.Horhold: *Patent DD*, (1993).
[21] V.A.Andryushina, T.S.Ctytsenko, T.G.Baklashova: *Patent RU* 2093518(1997).

In: Biotechnology, Agriculture and the Food Industry
Editor: G. E. Zaikov, pp. 157-162
ISBN 1-60021-040-6

Chapter 25

BACTERIA OF GENUS BACILLUS IN BIOLOGICAL CONTROL OF CULTIVAR PATHOGENS

T. V. Romanovskaya*, V. N. Kuptsov, A. E. Voronkova, N. V. Sverchkova and E. I. Kolomiets

Institute of Microbiology, National Academy of Sciences, Belarus

ABSTRACT

Bacterial strains of genus *Bacillus* – antagonists of cultivar pathogens were isolated. Physiological-biochemical properties and phytoprotective potential of isolates were examined. Prospects of using microbial antagonists for biological control of pathogens were demonstrated.

Key words: *antagonists, Bacillus, biological control, hyperparasitism, phytopathogens*

Current world practice envisages the concept of comprehensive plant biocontrol system aimed at replacing chemical pesticides with alternative ecologically safe biological analogs, including microbial agents. Recent investigations proved attractiveness of pathogen biological control using sporulating bacteria of genus *Bacillus* [1-4]. Representatives of this taxonomic group produce a broad spectrum of bioactive compounds, comprising enzymes, antibiotics, toxins, other metabolites with antimicrobial and insecticidal activity. Bacteria of genus *Bacillus* are inferior only to actinomycetes in range and amount of synthesized antibiotics, referred predominantly to polypeptides. Enzyme systems of bacteria incorporate a set of different enzymes providing the opportunity to utilize a spectrum of substrates and accounting for antimicrobial potential [5]. It seems natural therefore that research strategy focused on isolation and screening of highly active competitive strains-antagonists among sporulating bacteria for further application in biological control of most hazardous crop pathologies appears justified and extremely promising.

* Institute of Microbiology, National Academy of Sciences, Belarus 220141 Minsk, Kuprevich str., 2.E-mail:microbio@mbio.bas-net.by, fax: (017)264-47-66

Enrichment culture and membrane chamber techniques were used for strain isolation [6, 7] The latter method is most effective since it allows to obtain active antagonists against specific pathogens. To differentiate sporulating bacteria from species of other taxonomic groups soil samples before inoculation on agar media were subjected to thermal treatment (80^{0}C during 10 min). Preliminary screening was carried out by various methods [8[allowing to reveal diverse mechanisms of biotic interactions between isolated antagonists and phytopathogens.

Scrupulous screening procedure sorted out bacterial isolates 12A, K-3, B-37, 203 – promising agents for biological control of anthracnosis, brown and gray rot of lupine, fusarial, alternarial, rhizoctonial infections, phytophthorosis of vegetables (Table 1, 2). In the course of mixed incubation on agar media the examined strains produce specific zones of overgrowth on colonies of phytopathogenic fungi and bacteria, determined by hyperparasitic type of antagonism between microorganisms. It follows from analysis of antimicrobial range that despite the ability to suppress development of all phytopathogenic microorganisms, the tested antagonists display certain specificity of action. For instance, isolate 12A proved most effective against pathologies of vegetable cultures – rhizoctonial rot (*(R. solani 101 5D),* fusarial rot *(F. oxysporum P1)*, alternarial rot *(A. brassica A1),* gray rot (*B. cinerea B1),* whereas most active antagonists of lupine pathogens responsible for anthracnosis *(C. gloesporioides Л-2)*, brown spot (*P. setosa Л-3)* , gray rot (*B. cinerea Л-1)* are represented by bacterial strains *203, K-3* and *B-37.*

Table 1.Antifungal activity of tested microbial strains (wells technique)

Strain	Diameter of phytopathogen growth inhibition zones[1], mm						
	1	2	3	4	5	6	7
B. subtilis 12A	45,0±1,0	41,0±0,5	37,5±1,0	43,0±1,5	40,0±1,5	31,3±2,0	32,5±1,5
B. subtilis B-37	33,5±2,0	30,5±1,0	28,5±2,0	24,0±1,0	45,0±1,4	55,0±1,3	34,3±0,6
B. subtilis K-3	40,0±1,5	38,0±1,5	32,0±1,0	26,5±1,0	29,0±1,0	46,0±2,1	45,0±1,9
B. pumilus 203	30,0±1,0	37,0±1,0	33,5±1,5	42,5±1,0	37,5±1,5	24,3±1,0	32,0±109

Note: Pathogens of vegetable crops: 1 – R. solani 101 5D, 2 – B. cinerea B1, 3 - A. brassica A1, 4 – F. oxysporum P1, lupine pathogens: 5 - C. gloesporioides Л-2, 6 - P. setosa Л-3, 7 – B. cinerea Л-1

Table 2.Antibacterial activity of tested microbial strains (wells technique)

Strain	Diameter of phytopathogen growth inhibition zones[1], mm				
	1	2	3	4	5
B. subtilis 12A	21,5±1,0	39,5±1,0	43,0±1,5	51,0±2,0	38,5±1,0
B. subtilis B-37	19,5±0,5	28,5±1,5	26,3±1,0	30,0±1,5	17,0±0,5
B. pumilus 203	20,5±0,5	29,0±1,4	32,3±1,2	34,0±1,5	23,0±0,5

Note: Wheat and soya rot pathogens: 1 - P. syringae 239, 2 – P. syringae 280, pathogens of vegetable crops: 3 – X. campestris 33, 4 - X. beticola 8717, 5 – E. caratovora 662

Morphological and physiological-biochemical characteristics were studied for further identification of isolates. It was found that isolated bacteria form gram-positive motile rods sized 0.6-0.8x1.3-1μm with rounded ends, ellipsoid endospore is located in the central or subterminal part of the cell, spore bag is not enlarged. The strains yield on MPA spherical grayish colonies with smooth edge, 3-5 mm in diameter, slimy upraised in young cells or rough flat in old cultures. They produce brown pigment secreted into the medium on casein and glucose substrates. The afore-mentioned properties and analysis of differentiating characteristics summed up in Table 3, enable to affiliate isolates 12A, K-3, B-37 to species *Bacillus subtilis*, and strain 203 – to species *Bacillus pumilus*.

Table 3. Main diagnostic characteristics of isolated strains

Differentiating properties	12A, B-37, K-3	203
Cell diameter, μm	0,6-0,8×1,0-1,3	0,6-0,7×1,0-1,3
Spore bag	not enlarged	not enlarged
Motility	+	+
Gram staining	+	+
Anaerobic growth	–	-
Foges-Proskauer test	+	+
Reduction of nitrates to nitrites	+	–
Hydrolysis:		
starch	+	–
casein	+	+
tyrosine	–	–
Gelatin liquefaction	+	+
Acid generation from sugars:		
D-glucose	+	+
L-arabinose	±	+
D-xylose	+	±
D-mannitol	+	+
Utilization:		
citrate	+	±
propionate	–	-
Growth in 7% NaCl	+	±
Growth in 10% NaCl	–	–
Growth at temperature 50°C	±	±
Growth at temperature 55°C	–	–

Note: Symbol "+"- presence, "-"- absence, "±" - weak manifestation of the property

It was established that antifungal and antibacterial effects of *Bacillus* isolates - antagonists of phytopathogens damaging vegetable and grain-legume varieties were determined by produced metabolites (non-volatile and gaseous) and cells. Microscopic examination revealed inhibition of spore germination in phytopathogenic fungi under the impact of metabolites produced by tested antagonistic bacteria. In addition to physiological, the following and morphological changes were registered: mycelium vacuolization, formation of tumor-like structures on hyphas and growth tubes (Figure.). Shifts in phytopathogen morphology disrupt normal development cycle. The recorded destructive influence of antagonistic bacteria on phytopathogenic fungi is also supported by reports of other researchers [7].

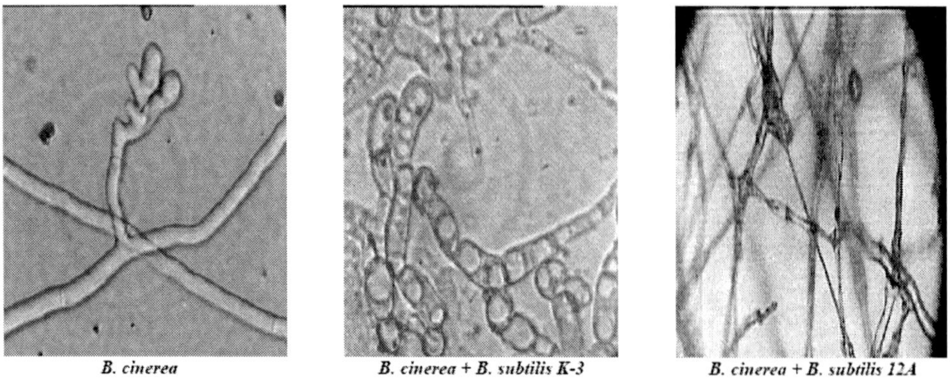

Figure. Influence of *B. subtilis K-3* and *B. subtilis 12A* metabolites onto mycelium growth of *B. cinerea*

Preliminary identification of antimicrobial metabolites produced by the isolated microbial species was conducted. It was found that substance referred to aliphatic ketones- 2,4-dimethyl-pentanone-3 was responsible for antibacterial action of *B. pumilus 203*, while antifungal activity of *B. subtilis K-3* was caused by synthesized polypeptide antibiotic.

Evaluation of phytoprotective potential of isolated bacteria-antagonists against lupine and vegetable crop diseases was performed in vitro and in vivo, It was shown that seed treatment with cultural liquid of bacteria *B. subtilis B-37* and *B. subtilis K-3* reduced damage induced by anthracnosis and gray rot in lupine seedlings (Table 4). Spraying of anthracnosis-infected lupine plants on field plots with cultural liquid of bacteria *B. subtilis B-37* inhibits further progress of the disease. Preventive effect was achieved with strain *B. subtilis K-3* on lupine trials against gray mould development.

Laboratory trials demonstrated that upon cabbage seed treatment with cultural liquid of bacteria *B. subtilis 12A* (titer 10^9 cells/ml) level of alternarial infection in seedlings fell 12-fold, bacterial infection – 3.5 fold, Phoma infection was completely suppressed (Table 5). Growth-stimulating action of tested bacteria was also detected. Germination index of treated seeds rose by 47 %, seedling length increased by 61 % as compared to the control. Monitoring of cabbage seedlings during vegetation period in field experiments established that seed treatment with suspension of bacteria *B. subtilis 12A* reduced black rot damage (caused by *Pythium debarianum Hesse*, *Olpidium brassicae Wor.*, *Rhizoctonia Aderholdii Kolosh, Fusarium sp.*) 2-fold and accelerated growth, indicated by increased height of the seedlings and enlarged plant assimilation area. Crude weight of 30 day plants grown from treated seeds tended to rise by 42.5 %, leaf surface by 12.7 % as compared to non-treated variant (Table 6). Economic efficiency of the procedure is demonstrated by extra yield of healthy standard seedlings from unit area. In our experiments with *B. subtilis 12A* this parameter equaled 17 %.

Table 4. Evaluation of phytoprotective properties of antagonistic bacteria

Type of seed treatment	Percentage of infected plants
Water	0
B. cinerea	100
B. cinerea + B. subtilis B-37	70
B. cinerea + B. subtilis K-3	0
B. cinerea + B. pumilus 203	70
C. gloesporioides	100
C. gloesporioides + B. subtilis B-37	15
C. gloesporioides + B. pumilus 203	50

Table 5. Biological activity of *B. subtilis 12A* following cabbage seed treatment (lab experiment, variety Belarusskay 85)

Parameters	Control	*B. subtilis 12A* cultural liquid	LSD_{05}
Germination power, %	86,2	99,5	-
Germination, %	87,0	95,0	-
Germination index	1,2	1,7	0,4
Seedling length, cm	2,6	4,2	0,7
Percentage of infected seedlings, %:			
bacterial rot	3.5	1,0	-
alternarial rot	6,0	0,5	-
phomosis	0,2	0,0	-

Table 6. Biological activity of *B. subtilis 12A* following cabbage seed treatment (field experiment, variety Belarusskay 85)

Parameters	Control	*B. subtilis 12A* cultural liquid	LSD_{05}
Black rot infection, %	5,7	2,8	-
Seedling height, cm	21,0	22,3	-
Crude weight,gr	80,0	114,0	31,0
Plant total leaf surface area, cm^2	21,2	23,9	2,3
Healthy seedling yield, number of plants per 1 m^2	148,4	173,6	-

The obtained findings support the assumption that the isolated strains may serve promising agents for pathogen biological control and lay the basis for engineering new bioproducts to protect vegetables and grain-legume crops.

REFERENCES

[1] B.Krebs, H.Junge, A. Ockhardt et al.: *Pesticide Science*. 37 (4), 427 (1993).

[2] D.Collins, C.Stevens, V.Khan, S.Nightengale: *Phytopathology*. 84 (10), 1114 (1994).

[3] Y.Huang, B.Wild, S.Morris: *Ann. Appl. Biol.* 120 (2), 367 (1992).

[4] M.Mari, M.Guizzardi, G. Platella: *Biol. Control.* 7 (1), 30 (1996).

[5] V. V. Smirnov: Sporulating aerobic bacteria – sources of bioactive compounds. *Nauk. dumka*, Kiev, 1983. 280 p. (in Russian).

[6] T.M. Lagutina, G.S.Muromtsev: *Application of combined chambers to monitor functioning microbial species and characterized their properties*. Leningrad, 1982. 18 p. (in Russian).

[7] N.S. Yegorova (Ed): *Guide to practical classes in microbiology*. MSU, Moscow, 1983. 215 p. (in Russian).

[8] A.M. Rosales, L. Thomashow, R.J. Cook, T.W. Mew: *Phytopathology.* 85, 1028 (1995).

In: Biotechnology, Agriculture and the Food Industry
Editor: G. E. Zaikov, pp. 163-167
ISBN 1-60021-040-6

Chapter 26

BIOTECHNOLOGY OF FAST BREEDING, IMPROVEMENT AND KEEPING OF COTTON VARIETIES

R. K. Shadmanov, L. B. Saranskaya and P. P. Paiziev

Institute of Genetics and Plants Experimental Biology, Tashkent Region, 702151, Kibrai District, p/o Ykori-Yuz, Uzbekistan

ABSTRACT

On the base of marking of such important traits as tolerance to wilt, salinity and drought as well earliness, fiber type and its quality, is shown a possibility of acceleration of breeding process and improvement of existing cotton varieties. The estimation of breeding materials takes place on the seed's level (genotypes) with use of marked enzymes and proteins. Using systematic analysis of seeds on all marked traits it is possible to reveal the most adaptable to unfavorable environmental factors genotypes with further their propagation. In the basis of this technology lies analysis of genotypes. A population, developed from such genotypes, has a high morpho-physiological homogeneity. On the base of the elaborated technologies we have obtained lines exceeding the initial sorts by many economic signs.

This technology is also applicable and to other types of plants.

Keywords: *cotton, genotypes, adaptation, wilt, enzymes, proteins, sorts.*

Success of genetic engineering and modern biotechnologies and their opportunities at reception of improved forms of plants do not cause doubts any more. Thus it shouldn't be denied a role of classical selection which continues and will remain predominating. Its combination with methods of the non-traditional selection based on biochemical and molecular markers can speed up this process two - three times and receive plants with desirable signs.

Having marked various signs of biochemical parameters of proteins and enzymes, it is possible to reveal the necessary genotypes arising in populations of varieties as a result of micro evolutionary processes and adaptation. However, morphologically they are often not expressed and selectively equivalent. For example the tolerant to wilt alleles are selectively equivalent to not tolerant, therefore their destiny depends on a case and corresponding frequencies will increase or decrease eventually in the casual way. It is in a population left for natural selection. But using an artificial selection of tolerant genotypes, we can support tolerance unlimited long time.

As the basic crop of our state is cotton, the creation of new varieties meeting all demands remains actual. Solving these problems, Classical selection is not always able to combine necessary signs in one variety, since it is conducted basically on high efficiency. Besides, process of common selection of cotton, based on hybridization and selection takes rather long time (up to 15 years).

On a background of a changing ecological situation and partly of the climatic factor connected with drying of Aral Sea pool, is felt more sharply necessity of fast creation of varieties with very short vegetative period, with maximal adaptability to factors of environment and with high economic parameters. Last achievements in the field of biotechnological researches open their great opportunities.

However, taking into account all of these research and achievements, it is necessary to note, that works on genetics engineering and many biotechnologies are rather expensive, labor-consuming, and it is enough long term subject. The main difference of our development is that we use natural genofunds of the plants population enriched during adaptation process. Under adaptation is usually understood long term process in which many generations are involved; along with that, adaptation may take place during the life of one organism.

Biochemical adaptation can be carried out at different levels of the metabolism, as well with the help of replacement of amino acids in primary sequence of enzyme, and with the help of change of isoenzymes balance, which causing the proteins polymorphism.

One of the major factors causing genetic polymorphism of natural populations including varieties is, as it has been mentioned, constantly happening micro evolutionary processes, speeded up action of both external and artificial factors of the environment. It leads to occurrence in population genetically heterogeneous individuals that finds reflection in distinctions of members of the population by inherited the diversified signs: morphological, physiological, biochemical, qualitative and quantitative. But physiologically and biochemical adaptive biotypes morphologically may not be shown. Therefore, selection by the phenotype of plants, especially in relation to pathogens, not always answers to the desirable or it is very long because of the factor of accident. Besides, it is necessary to take into account that microorganisms also adapt. Therefore frequency of biotypes occurrence on the basis of tolerance of long time used varieties is rather low because of occurrence of adaptive races of pathogens. But biotypes which have arisen as a result of adaptation, keep the high tolerance to pathogens much longer. It concerns to biotypes with other signs or possessing a complex of signs. Our technology, based on identification of such biotypes (genotypes) on marked signs allows very quickly, already at a level of seeds, to do their selection on many signs individually or on their majority in a genotype. Thus each sign was marked by enzymes or separate proteins of seeds which had direct or indirect relation to shown signs. As objectivity and reality of such kinds of markers, we show the characteristic of research deducing biochemical parameters to sign of tolerance to wilt disease.

Penetration of pathogen in a plant occurs in the initial stage of germination. Thus it should pass through epidermal layer of cells of seed covers or root. At presence of active enzyme systems in this layer, occur intensive oxidizing processes. Formed products of oxidation possess stronger fungicide activity in comparison with initial substances . Being allocated for a surface of epidermal layer in places of contact with an infection, they form a powerful chemical barrier.

Therefore, many researchers observed, but could not explain mass fungi settling on the surfaces of a root and its insignificant penetration into tissue of the plants with low disease and practically full its absence on roots of infected plants and with high content of the infection in their tissue . Meeting on the way of introduction a number of cells with the higher activity, the infection eliminate (immune genotypes) or gets into vascular system of plant weakened or in insignificant amount. That does not influence essential on processes of normal development (tolerant genotypes). We have shown, that for resistance to wilt (it is possible also to other fungi) to defeat answer two non mitochondrial chain of breath, not developing ATP and final oxygenizes which are phenol - and ascorbatoxidase . Presence of these chains metabolism in plants was known long time, but their role in ability to live of plants was not clear. We found out that activity of these systems in tissues of plants determines degree of stability. A role of system polyphenol-polyphenoleoxidases in plants immunity is also known. But it could not explain : why at presence of this mechanism and the sufficient content of phenols in plants, wilt disease is rather high, especially in cotton plants (gossypol). There were even attempts of additional processing of plants by gossypol. But they also had no success. The reason was that very high activity of polyphenoloxidase has not been connected with immunity but with mitochondria's respiratory system, causing development of ATP. Nonmitochondrial system of oxidizes could be very much feebly marked and even to be absent. Naturally, the mechanism of oxidation of phenols in process of penetration of an infection didn't work. Analyzing tissues of plants in which the content of mitochondria is minimum, we found out, that not all genotypes reveal presence of nonmitochondrial systems of breath. Resistant genotypes always have them in amount, which provides immunity. Also it has been shown, that phenoloxidase joins at the initial period of seeds germination, and ascorbatoxidase works during the budding and flowering period. The posterity of the genotypes selected in accordance with these signs, possesses a high degree of tolerance to pathogens in comparison with the initial material. Periodically repeating this reception, it is possible to prolong preservation of a variety for many years. Advantages of this method of increase tolerances of cotton to adverse factors are obvious. First of all it is exception of accident. Besides dealing with seeds, as the selection is conducted at a level of seeds, work can be conducted within a year, making multiple copies of plants during the winter period in hothouses. Besides it is essential economy of time, that is in improvement of a sort or selection process on rearing of new sorts of great importance. And, at last, similar reception is applicable to other cultures. Similarly were marked the traits of earliness, high oil content, natural early leaf fall, and resistance to salinity and to drought. As a result of selection of the genotypes according to these two parameters have been received cotton lines considerably surpassing the initial material on required signs.

TECHNIQUE

Two industrial varieties AN Bayaut-2 and C-4727 served as initial material for research. Varieties are strictly reigned in a zone with low infection of wilt in the ground. The micro quantity of germs was used for the analysis of seeds (genotypes) which was exposed to electrophoretic analysis for revealing of marked enzymes in it. At detection of necessary parameters, the rest of isolated germ, has been growing in laboratory conditions and then as sprouts it was transferred to a field. Further in field conditions were made multiple copies of all genotypes selected on similar parameters. As a result it turned out morphological completely leveled linear material ready to subsequent duplication and transfer for testing.

Results of Research

The results of research are submitted in tables 1-3.

Table 1.Economic characteristic of Lines (Features)(Year 2003)

Materials	Vegetative Period (days)	Number of bolls on a bush	One boll weight (gr.)	Fiber output (%)	Fiber length (mm)	1000 seeds Weight (gr.)
Namangan-77 (Standard)	108	13,2	5,3	38,2	32,7	115,0
Bayaut-2 (initial form)	114	12,1	5,5	35,7	32,1	115,0
Line L-19	108	15,2	7,4	38,7	33-34	141,0
Line L-12	108	14,0	7,0	38,5	33,0-34,0	139,5
C-4727 (initial form)	113	13,1	6,0	36,2	32,4	125,0
Line L-3	110	23,5	6,6	38,6	34,7	125,0

Table 2.Wilt disease degree (Year 2003)

Materials	Number of registered plants	Wilt diseased	
		Number of Plants	%
Namangan-77 (Standart)	320	64	20,0
Bayaut-2 (initial form)	150	94	62,7
Line L-12	250	9	3,6
C-4727 (initial form)	200	170	85,0
Line L-3	240	11	4,6

Table 3. Oil content in seeds of the Cotton sorts and new Lines (Year 2003)

Sorts and Lines	% of oil in a seed	% oil in a nucleus
Namangan-77	19,0	31,3
БАЯУТ-2	20,4	35,9
Line L-19	25,6	39,8
Line L-12	25,1	39,5
C-4727	22,9	37,6
Line L-3	24,5	39,9

According to the Tables all parameters of received Cotton Lines significantly exceed the control sort's characteristics.

Weight of row cotton on the first and second position on simpodium and from the first up to the last simpodium of the allocated lines was practically identical and very high - 7.5-8.0g. Row cotton weight on the third position of simpodium was lower - 6,5g. However, average weight on simpodien and places on a bush amounts as a whole almost to 7g. Simple calculation shows, that, even at presence of 8 - 10 boxes on a bush and number of plants on hectare up to 90-100 thousand, the outcome can exceed 6.0 t. Also taking into account high uniformity and short time of reception of such materials, it is possible to realize the importance of this technology for practice.

REFERENCE

[1] Shadmanov R.K. et al. –*China Cottons*, 23(8),1996.

[2] Shadmanov R.K. - In *Turk World Symposium* , p.251-254, 1999.

[3] Abdukarimov A.A. - International Conference of Science and Technology Shadmanov R.K. et al. for *Managing Plant Genetic Diversity* in 21 Century.Malaysia, Kuala Lumpur, 2000.

[4] Abdukarimov A. A. - *Бюлл.ГКНТ РУз*,№3-4,с.24-43, 2000.

[5] Shadmanov R.K. Shadmanov R.K. - *World Cotton Research Conference*-3, Cape Town, South Africa,2003- Saranskaya L.B

In: Biotechnology, Agriculture and the Food Industry
Editor: G. E. Zaikov, pp. 169-172

ISBN 1-60021-040-6

Chapter 27

SELECTION RESISTANT CALLUS LINES RAPE (BRASSICA NAPUS L.) TO STRESSFUL FACTORS

I. D. Sytnik and O. L. Klyachenko*
National Agriculture University, Kyiv 03041, Ukraine

ABSTRACT

Influence on body height callus tissues of rape osmotic and cold a stress was investigated. As reacting factors used PEG-6000, mannitol and low temperature. As a result of researches were selected drought- and coldresistant callus lines of rape and plants-regenerates from them are received.

Key words: *rape, osmotic and cold a stress, callus tissue, polyethylene glycol, mannitol.*

Rape as the source of alimentary oil and fodder protein takes the important place in the decision of a problem of a saturation of the market the foodstuffs. However sorts of domestic selection which are raised on Ukraine I do not fully comply with requirements of effecting behind quality of oil, and foreign - are poorly adapted to soil-climatic environments of Ukraine.

Resistance of plants — geneticaled caus attribute. But as he is potential shows only in environments of a stress. Therefore the success in making resistant sorts depends on presence of effective methods of diagnostics and sources of high resistance of a starting material for selection. The special interest for environments of Ukraine represents increase of resistance rape to unfavorable factors of environment, namely to a drought and low temperatures. For today works on cellular selection resistant against a drought and low temperatures of agricultural plants at a level callus, cellular suspensions and the whole plants [1-4] are known.

The purpose of our work was selection and reception callus lines of rape resistant against low temperatures and an osmotic stress.

* National Agriculture University, 15, Heroev Oborony St., Kyiv 03041, Ukraine e-mail: Klyachenko@ukrpost.net

Procedure

The object was the cellular suspension of various genotypes of rape. Tootain callus lines if sugar beet we used nutrient medium of Murasige-Scoog: 2,0 mg/l NAA + 0,4 mg/l BAP + 2,5 mg/l of an ascorbic acid. Callus is received after 2-3 weeks of cultivation in diffused light on medium with 24-25°C temperature. Suspension crop was cultivated during 14 days in darkness, in liquid nutrient medium (2-3 g callus tissues in 60 ml of medium) at constant mixing on a shaker (90-100 rev/min) and temperature 25-26°C.

Research carried out under the circuit of classical selection. With the purpose of imitation in vitro effect of the deaquation accompanying a drought, was applied nutrient medium, which was supplemented osmotically active materials — polyethylene glycol PEG-6000 and mannitol [4, 6]. For reception of coldresistant plants callus tissue of rape was maintained at temperature +1°C during 6 hours. A survival rate callus ines was defined on an accretion callus tissue [7].

Results and Discussion

On Ukraine the significant area of arable lands is placed in region of insufficient and unstable humidification. To lower risk of short-reception of production it is possible due to cultivation of drought-resistant sorts. Development of methods of selection of drought-resistant forms in culture in vitroò an opportunity to carry out scale selections and tests of the best selection samples.

For studying influence of water deficiency on callus culture of rape in nutrient medium was brought PEG-6000 in concentrations of 5-25 % and 15-20 % mannitol which represent osmotically awake materials not inpouring into a cell. Taking into account unequal ability of various genotypes to persist at introduction in structure of nutrient medium of osmotically awake materials, in the researches we used genetical various stuff - 23 genotypes (tab. 1).

As a result of the researches was established, that at selection of drought-resistant forms of rape behind a parameter of osmotic activity of an organism concentration PEG-6000 and 17 % mannitol is optimum 12 %.

The cause of a stress at plants was low temperature, it is especial concerns spring rape which due to difference of temperatures in the early spring very much freezes slightly. Receiving of initial selective material, resistant to extreme temperatures, is possible at use of methods of cellular selection. The cold stress caused plants the big temperature range: from 0 to +8°C. In the researches was used callus tissues was received from cotyledonous leaves and meristems of a caulis of various genotypes summer rape. A degree cold damages was defined on an accretion callus tissues after refrigerating. Compared a degree cold damages of lines of the callus lines received from various explants, sustained at temperature +1°C during 6 hours. For this purpose was took explants with known weight (50 mg) from refrigerated and control callus, was placed them on nutrient medium and raised during 14-28 day then was weighed. Accretion of crude mass was expressed in percentage to the initial data. The highest cold constancy was marked at callus cultures received of cotyledonous leaves, and the lowest – at callus meristem parentages (tab. 2).

Definition of an accretion callus after refrigerating enables to make the assumption of duration of conservation of unfavorable influence of the lowered temperatures on body height of cells. As a result of researches were selected frigostable callus lines of sorts spring rape Aira, Likolli and plants-regenerates which was used as a valuable starting material at receiving frigostable sorts rape are received from them.

Table 1. Influence PEG-6000 and mannitol on an accretion of callus tissues of various sort rape

Sorts	Concentration in nutrient medium, %									
	PEG 6000						Mannitol			
	Control with out PEG and mannitol	5	9	12	16	20	12	15	17	20
Mariya	94,8	88,3	63,5	40,1	22,4	10,8	72,8	52,8	34,5	18,9
Oksamit	93,9	86,4	64,1	38,2	21,3	8,4	70,5	51,2	30,9	16,5
Aira	94,3	87,2	63,0	38,7	22,1	9,2	70,9	51,6	31,8	16,8
Otma	95,6	88,9	65,3	41,6	10,9	10,9	74,6	54,3	38,2	20,1
Sirius	96,1	89,2	67,2	42,0	11,0	11,0	75,3	53,9	37,4	20,3
Sriblyastyj	95,3	89,1	69,2	42,9	11,2	11,2	75,4	55,3	39,6	21,1
Zemir1	94,4	84,3	60,5	34,2	6,4	6,4	69,4	48,8	30,2	16,4
Likolli	92,9	88,0	66,4	36,8	6,8	6,8	70,8	50,7	33,8	17,8
Kwantum	94,8	87,5	62,5	35,7	6,3	6,3	71,3	49,6	31,6	16,9
Forte	97,3	83,6	55,6	31,8	5,2	5,2	67,4	45,4	27,9	14,1
Kosmol	96,5	85,9	60,2	32,9	5,3	5,3	70,2	47,3	28,4	15,3
Kawiar	90,4	86,3	64,1	33,4	5,5	5,5	71,9	50,2	30,1	16,8
Hajola	96,2	80,8	54,9	30,0	4,0	4,0	64,5	42,1	24,8	12,6
Lizone	92,8	88,5	61,8	32,4	3,9	3,9	69,3	45,8	25,9	13,2
Liga 1	95,3	84,4	57,7	28,6	3,5	3,5	66,7	44,5	24,4	12,1
Otaman	94,2	79,4	54,3	26,4	3,2	3,2	63,8	40,7	20,8	10,8
Arion	95,9	79,1	50,8	24,5	3,6	3,6	60,2	38,2	20,2	10,7
Titan	94,1	81,2	51,6	24,9	3,0	3,0	61,9	39,6	21,6	11,5
Iris	92,5	83,8	52,8	25,3	3,1	3,1	62,6	38,0	21,3	11,6
Mikitheschkiy	90,3	77,6	49,0	20,1	2,9	2,9	58,1	36,3	18,4	9,7
Kalinivskiy	91,6	75,3	47,3	19,4	2,9	2,9	56,3	33,9	14,6	8,1
Shpat	91,9	70,2	44,6	18,2	2,6	2,6	50,9	30,1	13,0	8,1
Oredeg 2	92,8	69,7	42,1	18,0	2,8	2,8	49,1	29,4	11,5	6,6

Table 2. Influence of 6-hour refrigerating at temperature +1°C on an accretion callus masses of summer rape a various parentage

Sort, explant	Masse after 14 days, mg		Masse after 28 days, mg		Accretion, %	
	control	refrige-rating	control	refrige-rating	control	refrige-rating
Mariya						
cotyledonous leaves	73,0±2,1	57,2±1,3	91,6±3,1	66,8±1,7	183,2±4,2	133,6±3,6
meristem	76,2±2,3	53,1±1,1	92,7±1,7	60,41±2,3	185,4±3,7	120,8±2,8
Aira						
cotyledonous leaves	78,1±3,1	59,3±1,4	95,8±3,4	71,6±3,3	191,6±4,0	143,2±3,8
meristem	81,4±3,3	55,4±1,0	96,9±2,9	62,7±2,7	193,8±3,4	125,4±2,6
Kawiar						
cotyledonous leaves	62,3±2,2	48,7±0,9	80,5±2,5	66,5±1,6	161,4±4,4	113,0±1,8
meristem	66,1±2,4	47,1±0,8	83,2±2,7	52,9±1,8	166,4±4,2	105,8±2,4
Kosmol						
cotyledonous leaves	60,5±1,9	47,3±1,2	79,4±1,8	58,1±2,9	158,8±4,7	116,2±3,2
meristem	63,4±2,1	45,4±1,9	82,6±3,5	82,6±3,5	165,2±4,5	104,0±2,7
Likolli						
cotyledonous leaves	75,8±2,5	59,5±2,6	94,0±4,1	70,8±3,8	188,0±4,9	141,6±3,7
meristem	80,3±2,8	52,4±2,3	98,6±3,9	60,9±3,6	197,2±4,5	121,8±2,5

REFERENCES

[1] V.A. Sidorov: Biotechnology of plants. *Cells selection*. Kiev, 1990. 280 c. (in Russian).

[2] R.A. Bressan, P.M. Hasegana, A.K. Handa: *Plant Sci Lett*., (21), 23 (1981).

[3] J.W. Heyser, M.W. Nabors: *Plant Physiol*., (63), 63 (1979).

[4] A.S. Lukatkin: *News AS*., (3), 304 (1998). R. H. Smith, S. Bhascaran, F. R. Miller: In vitro Cel. *And Dev. Biol*., 21 (10), 541 (1985).

[5] C.T. Harms, J.J.Oertli: *Plant Physiol*., (12), 29 (1985).

[6] L.A. Kucherenko, R.P. Maddumage, J.L. Gughov: *Agricultural biology*. (3), 84 (1991).

INDEX

A

B

C

D

E

F

G

H

I

N

O

P

Q

R

S

T